WHEN WE ARE NO MORE

WHEN WE ARE NO MORE

*How Digital Memory Is
Shaping Our Future*

ABBY SMITH RUMSEY

BLOOMSBURY PRESS

NEW YORK · LONDON · OXFORD · NEW DELHI · SYDNEY

Bloomsbury Press
An imprint of Bloomsbury Publishing Plc

1385 Broadway   50 Bedford Square
New York   London
NY 10018   WC1B 3DP
USA   UK

www.bloomsbury.com

BLOOMSBURY and the Diana logo are trademarks of Bloomsbury Publishing Plc

First published 2016

ISBN: HB: 978-1-62040-802-5
         ePub: 978-1-62040-803-2

LIBRARY OF CONGRESS CATALOGING-IN-PUBLICATION DATA HAS BEEN APPLIED FOR

British Library Cataloguing-in-Publication Data
A catalogue record for this book is available from the British Library.

2 4 6 8 10 9 7 5 3 1

Typeset by RefineCatch Limited, Bungay, Suffolk
Printed and bound in USA by Berryville Graphics Inc., Berryville, Virginia

To find out more about our authors and books visit www.bloomsbury.com.
Here you will find extracts, author interviews, details of forthcoming events
and the option to sign up for our newsletters.

Bloomsbury books may be purchased for business or promotional use. For information
on bulk purchases please contact Macmillan Corporate and Premium Sales Department
at specialmarkets@macmillan.com.

for
David Rumsey
Hero *of* Time *and* Space

CONTENTS

PART ONE

WHERE WE COME FROM

I imagine the earth when I am no more:
Nothing happens, no loss, it's still a strange pageant,
Women's dresses, dewy lilacs, a song in the valley.
Yet the books will be there on the shelves, well born,
Derived from people, but also from radiance, heights.
　　—CZESLAW MILOSZ, "AND YET THE BOOKS," 1986

Chapter One

Memory on Display

O VER FORTY THOUSAND YEARS AGO, humans discovered how to cheat death. They transferred their thoughts, feelings, dreams, fears, and hopes to physical materials that did not die. They painted on the walls of caves, carved animal bones, and sculpted stones that carried their mental and spiritual lives into the future. Over generations we have created sophisticated technologies for outsourcing the contents of our minds to ever more durable, compact, and portable objects. Each breakthrough in recording technology, from the creation of clay tablets six thousand years ago to the invention of papyrus scrolls, printing, photography, audio recording, and now ultracompact, portable, and extremely fragile digital media, has added to the vast stores of knowledge that hold the key to our success as a species. In the digital age we are dramatically expanding our capacity to record information, freeing us to pursue our curiosity at will and seek answers to ever more ambitious questions.

But every once in a while, we outsmart ourselves, and we have to scramble to catch up with our inventions. This is such a moment. The carrying capacity of our memory systems is falling dramatically behind our capacity to generate information. Since the creation of the World Wide Web in the 1990s and the growth of social media in the last decade, we feel increasingly overwhelmed by information. At the same time, we are intrigued—if not downright infatuated—with the power and promise of this abundance. We demand more and more—Big and Bigger Data. Yet it seems the more information we have, the less we feel in control of what we know. How do we catch up with ourselves now?

This is not the first time humanity has felt overwhelmed by the riches created by our ingenious inventions. Every innovation in information technology, going back to ancient Mesopotamians' invention of cuneiform tablets, precipitates a period of overproduction, an information inflation that overpowers our ability to manage what we produce. Having more knowledge than we know what to do with while still eager to acquire more is simply part of the human condition, a product of our native curiosity.

But this moment is different in quality as well as quantity. We can no longer rely on the skills we have honed over millennia to manage our knowledge by managing physical objects, be they papyrus scrolls or paperback books. Instead, we must learn to master electrical grids, computer code, and the massive machines that create, store, and read our memory for us. What this mastery looks like and how we achieve it is today's frontier of knowledge.

The digital landscape before us is largely unmapped, terra incognita that we can only know by entering into it and exploring. Fortunately, vast as the unknown territory may be, digital technology itself helps speed communication of new knowledge between those rushing ahead to explore the unknown and those traveling at a slower pace who are settling the new landscape and making it productive. As the frontier retreats quickly before us, we can already see that our age-old understanding of humanity's collective memory as something fixed to durable objects and constrained by the limits of time and space is obsolete. Digital memory is ubiquitous yet unimaginably fragile, limitless in scope yet inherently unstable. Mastery of digital memory means grappling with its vulnerabilities as well as developing its strengths. We will explore both as we examine the future of memory in the digital age.

The consequences of going digital for the future of human memory came into sharp focus for me in 1997, while leading a team of curators at the Library of Congress to assemble a comprehensive exhibition of its collections for the first time in living memory. The library had just acquired its one hundred millionth item. From this abundance we were to select several hundred items that would tell the two-hundred-year story of the Library of Congress and, by extension, the American people. We had much—too much—to choose from. Home to the United States Copyright Office and faithful to its founder Thomas Jefferson's vision of creating a universal and comprehensive collection of human knowledge, the library has records in virtually every medium capable of carrying information, from rice paper and palm leaves to mimeographed sheets and onionskin paper, whalebones and deer hides, audio wax cylinders, early television kinescopes, silent movies on nitrate film, maps on vellum, photographic negatives on glass plates the size of tabletops— and, of course, computer code on tape, floppy disks, and hard drives.

It was remarkably easy to select several hundred objects out of one hundred million because each object tells a tale. To tell the story of how the Republic was born, for example, we displayed the Rough Draft of the Declaration of Independence, crafted over a few days in July 1776 by Thomas Jefferson and edited by Benjamin Franklin, John Adams, Roger Sherman, and Robert Livingston. It is written in the eminently legible hand of Thomas Jefferson. Yet several passages are boldly struck through with lines of heavy black ink and emended with the changes made by Adams and Franklin.

The sight of Jefferson's venerated text so vividly edited always draws people up short. They are startled to see that the most famous phrase in this most famous document—"we hold these truths to be self-evident, that all men are created equal"—is not what Jefferson wrote. He wrote that the truths are "sacred and undeniable." The words we know so well today are in fact a correction suggested by Benjamin Franklin. The jarring yet oddly familiar sight of the Declaration of Independence in full Track Changes mode makes self-evident the disagreements among the Founders and the compromises they agreed on. The original document renders the past strangely new—the events dramatic, the motives of the actors complicated, the conclusion unpredictable.

Historians continue to mine the Rough Draft's four pages of tangible evidence for clues to the early stages of the colonial rebellion. As a historian, I was familiar with the excitement of working with original documents. I also knew how stirring—at times emotional—it is to work directly with originals. A physical connection between the present and past is wondrously forged through the medium of time-stained paper. Yet what I remember most vividly is the impact of the Rough Draft on tourists. Many of the visitors had stopped by the library simply as one more station on a whirlwind circuit of the capital. They were often tired and hot and not keen on history in the best of circumstances. But this was different. They would grow quiet as they approached the exhibit case. They lowered their heads toward the glass, focused on lines of text struck through to make out the words scribbled between lines, and began to grasp what they were looking at. Their reactions were visceral. Even dimly lit and safely encased in bulletproof glass, the Rough Draft emanates an aura of the "sacred and undeniable."

It was then that I started to think seriously about the future of memory in the digital age—though worry is the more accurate word. What would my successor show in two hundred years' time—or even fifty years? How would people feel that distinctive visceral connection with people from the past if the past had no undeniable physical presence? What we displayed in 1997 had withstood the test of time. It was already self-evident that there would be no test of time for digital information. At that time, web pages lasted an average of forty-four days before changing or disappearing altogether. We seemed to be moving at breakneck speed from a knowledge economy of relative scarcity of output to one of limitless abundance. By latest count in 2015, the Library of Congress had well over 160 million items, already a startling increase over the 100 million it counted in 1997. But relative to what circulates on the web, its collections could be described as if not scarce at least tractable. Engineers at work on building the largest radio telescope in the world, the Square Kilometre Array, estimate that when the telescope is up and running, it will produce "up to one exabyte (10^{18} bytes) of data per day, roughly the amount handled by the entire Internet in 2000." And the web itself grows inexorably. One data-storage company estimates that worldwide, web data are growing at a rate that jumped from 2.7 billion terabytes in 2012 to 8 billion terabytes in 2015. But nobody really knows—or even agrees how we should be counting bits.

How are we to keep from being drowned in the data deluge? In the past, the materials used for writing, the human labor of copying, the costs of disseminating and providing access to books, atlases, photographs, films, and recorded sound were very high. These costs imposed limits on the rate of production, in effect filtering what knowledge and creative expression were accessible and to whom. The expense of maintaining vast and redundant stores of physical artifacts meant it was costly to collect them and invest in their long-term access. The question had always been: "What can we afford to save?"

Now, suddenly, those filters are gone and information travels at the speed of electrons, virtually free of friction. Now everyone with a computer can publish their own book, release their own movie, stream their own music, and distribute what is on their hard drive or smartphone across the globe instantaneously. The question today is: "What can we afford to lose?"

Though this seems a daunting question, we have a lot of information from the past about how people have made these choices before, in other periods of information inflation—and there have been many. They routinely follow every innovation in recording technologies. It happened when Sumerians first invented writing to store information about grain harvests and found themselves puzzled by where to put so many clay tablets so they would be safe from damage or theft but also easy to get to when needed. It happened when Europeans invented printing and the marketplace filled up with competing and contradictory versions of canonical texts like the Bible. It happened again when we created audio recordings on platters that would break if handled roughly and moving images on nitrate film stock that could ignite and blow up, even in the absence of oxygen. Each innovation prompted a rethink about how to use these astonishing new powers of communication, each full of unknown potentials that could be uncovered only through experimentation. And each advance required a very costly retool of the information infrastructure already in place. Creators, publishers, librarians, and archivists all scrambled to catch up. But it was always worth the price, no matter how high it seemed at the time, because we gained the freedom to reimagine our collective memory, confident that we could capture so much more of the human experience.

This growing body of shared knowledge and know-how decisively shapes our fate as a species, distinct from all others. Over generations, as we perfected the technologies of recording and created more resilient and compact media to hold our knowledge, we gained dominion over the planet. Our culture and technologies are the ultimate power tool, enabling adaptive strategies that far outpace the strictly biological evolution other species must make do with. Yet quite abruptly and without warning, at the beginning of the twenty-first century we embarked on a vast natural experiment, rendering obsolete our forty-thousand-year project to cheat death by using objects to hold the contents of our minds. Gone is the promise of preserving knowledge forever. We are replacing books, maps, and audiovisual recordings with computer code that is less stable than human memory itself. Code is rapidly overwritten or rendered obsolete by new code. Digital data are completely dependent on machines to render them accessible to human perception. In turn, those machines are completely dependent on uninterrupted supplies of energy to run the server farms that store and serve digital data.

How do we guarantee that this uncontrolled experiment with human memory will turn out well for us? In our search for the answers, we will look back, exploring the history of how we have mastered the challenges of information inflation before. And we will look inward, into the human mind, to gain often surprising and frankly counterintuitive insights into how the brain's natural filtering systems manage to determine what information to save and what to dump without any help at all from our conscious minds. Both historical experience and contemporary science provide insights critical for sustaining the collective memory of humanity and managing our own personal digital archives.

WHAT IS THE BIG IDEA?

Two reigning misconceptions stand in the way of a happy ending to our experiment in reimagining memory for an economy of digital abundance. First is the notion that today's abundance is a new phenomenon, unseen in human history, which began with computers and is driven by technology. This is like blaming a mirror for the blemish you see on your cheek. Technology is an instrument of the human will, not vice versa. True, the pace of information production has accelerated, and the fact that computers make perfect copies so easily certainly accounts for the growth in redundant information that we store. It is effortless and seemingly inconsequential to hit Forward and further inflate the ballooning data universe. But the current information inflation began not in the 1990s, when the Internet was opened to commerce, nor in the 1940s, when computers were built by the military. It began in the first half of the nineteenth century. And it was not a technical innovation that set us on the present course, but an idea. That was the radically transformative idea that the universe and all that exists is no more and no less than the material effect of material causes.

This idea, known in philosophy as materialism, itself is ancient. It was central to the thinking of the Greek Democritus (ca. 460– ca. 370 B.C.), immortalized in a poem by Lucretius (ca. 99 B.C–ca. 55 B.C.) called *On the Nature of Things* (*De rerum natura*), and can be found in ancient Indian and Chinese philosophies. But in the hands of Western men of science (and they were mostly men), the view of matter as both cause and effect did not serve the sole purpose of understanding the world as philosophers. They sought new knowledge to master Nature's secrets and their political counterparts sought to use that knowledge to change the world. And so, by the 1830s, the great hunt for physical evidence was on. The rapid invention of tools of investigation resulted in a proliferation of new information technologies. From the daguerreotype invented in 1838 to the powerful imaging technology at the heart of the Large Hadron Collider that detected traces of a new subatomic particle in 2013, our information technologies all derive from the single insight that matter records the history of the universe because it is a slow, cold form of information. The universe writes its own autobiography in atoms. The evolution of our collective memory from Paleolithic cave paintings to the World Wide Web is the story of how and why this idea of matter as memory took hold and what it means for us today.

Culture evolves in fits and starts. History is studded with false promises and dead ends, experiments that work for a while then prove unfit as circumstances change. But there are also moments of rapid change, inflection points when forces coalesce to accelerate and alter the trajectory of events. Four inflection points in particular precede and enable the scientific advances of the nineteenth century that inaugurated today's information inflation: (1) the development of writing in Mesopotamia for administrative and business purposes, together with professional management of the collections; (2) the ancient Greeks' development of libraries as sites for the cultivation of knowledge for its own sake; (3) the Renaissance recovery of Greek and Roman writings and the invention of movable type, which together helped to propel the West into the modern age; and (4) the Enlightenment of the eighteenth century, which refashioned knowledge into an action verb—progress— and expanded the responsibilities of the state to ensure access to information.

These inflection points all lead up to the critical moment around the turn of the eighteenth century when some curious geologists discovered that rocks are like clocks. Properly read, they could be used to tell the time of the Earth. And the Earth turned out to be far older than people thought. That was the moment when science moved from the Age of Reason to the present Age of Matter and the great quest for physical evidence about all that exists began. The digital era is merely the most current installment in the unfolding saga of our desire to know more about the world and ourselves. The origins of today's global information economy of abundance lie here, at this inflection point in the history of Western thought. For it is the West that created the global inscription used today—the digital code that travels on worldwide networks.

WHAT IS MEMORY?

The second misconception is our antiquated view of memory itself. The computer is not an accurate model for the brain. Scientists now understand natural memory—the kind that hedgehogs and humans have, as opposed to the artificial kind we use for storing information like books and silicon chips—is the primary mechanism animals rely on to adapt to their environment. Memory is the entire repertoire of knowledge an animal acquires in its lifetime for the purpose of survival in an ever-changing world—essentially everything it knows that does not come preprogrammed with its DNA. Given the complexity of the world, memory takes a less-is-more approach. It is sparing, even provident in its focus on information that may come in handy later. Like a traveler packing for a week trying to squeeze all necessities into an overnight bag, the brain compacts big loads of information into small spaces by combining and compressing similar information through elaborate networks of association.

We keep our mental model of the world up to date by learning new things. Fortunately, our memory is seldom really fixed and unchangeable. If it were, that would put us in extreme peril—unless the world were to suddenly stop, become fixed and unchangeable too. We must be as adept at forgetting what is no longer true or useful as we are at remembering what is valuable and necessary. As we take on new roles and responsibilities in life, such as parent, partner, worker, or citizen, we shed old ones—child, student, or dependent. Like muscles, memories weaken with time when they are not used. Just as in the art of packing, in which what we leave out is as important as what we put in the bag, so too does the art of memory rely on the art of forgetting.

What this means for the digital age is that data is not knowledge, and data storage is not memory. We use technology to accumulate facts about the natural and social worlds. But facts are only incidental to memory. They sometimes even get in the way of thoughtful concentration and problem solving. It is the ability for information to be useful both now and in the future that counts. And it is our emotions that tell what is valuable for our survival and well-being. When distracted—for example, by too many bright shiny things and noisy bleeping devices—we are not able to learn or develop strong reusable memories. We fail to build the vital repertoire of knowledge and experience that may be of use to us in the future. And it is the future that is at stake. For memory is not about the past. It is about the future.

Human memory is unique because from the information stored in our brains we can summon not only things that did or do exist, but also things that *might* exist. From the contents of our past we can generate visions of the future. We know there is a past, a present, and a future, and in our heads we travel freely among these time zones. We know that generations of people were born and died before we came into existence, and that we, too, are destined to die. This deep temporal depth perception is unique in Nature. We engage in mental time travel, imagining possible future outcomes, or traveling backward in time to re-create how something in the present came to be this way and not that. The richer our memories, the greater our imaginative capacities. And our destiny as problem solvers lies in this conjectural thinking.

As we consider memory in the digital age, we will see how our personal memory is enhanced, and at times compromised, by the prodigious capacities and instantaneous gratifications of electronic information. We will also look at memory "at scale," as computer engineers would say, the collective memory of humanity as it has grown and at times shrunk over thousands of years. Collective memory—the full scope of human learning, a shared body of knowledge and know-how to which each of us contributes and from which each of us draws sustenance—is the creation of multiple generations across vastly diverse cultures. We will see how single acts of learning can be shared and accumulate over time to become knowledge, how generations have worked to secure that knowledge and share it across generations, and how each generation builds upon this foundation. Digital networks make our collective memory accessible across political and linguistic boundaries. Everyone with access to the Internet can turn personal memory and learning into shared knowledge, ensuring that the collective memory of humanity continues to be culturally diverse as it grows exponentially.

The past as well as the future of this collective memory is being fundamentally reshaped by digital technology. How do we guarantee that this uncontrolled experiment in human memory will turn out well? There are no guarantees, of course. But what happens is in our hands. We face critical decisions as a society and as individuals about how to rebuild memory systems and practices to suit an economy of information abundance. It is rare that any generation is called upon to shape so much of the world that future generations will inherit. It is my goal to deepen our understanding of memory's role in creating the future and to expand the imaginative possibilities for rebuilding memory systems for the digital age.

This is a little book about a big idea. It is not a book of predictions, for the future is unknowable. Nor is it a comprehensive history or analysis of cultural and biological memory. Instead, it is an exploration into new territories of memory, past and future. Like all travelers making their way in terra incognita, we have to take strategic detours in order to stay headed in the same direction. We must, with regret, sidestep some very interesting but ultimately distracting diversions along the way. For those who want to explore a topic we consider too briefly for their taste, I have included a list of the key resources I used in my research, as well as pointers to specific sources and additional commentary in the notes.

Our journey begins by looking behind us, into the deep past of human memory, to learn how we arrived at this pass. Along the way, we spend time at key inflection points with individuals whose stories exemplify how changing ideas about memory interact with technologies for sharing knowledge to expand human potentials. We survey the biology of memory, a field still in its infancy, to gain insights into the brain's natural filtering systems that capture what is valuable and dump the rest. We end in the present day to consider the personal, social, and cultural choices we face as we work to master the abundance of memory in the digital age.

CHAPTER TWO

How Curiosity Created Culture

Thoughts in the concrete are made of the same stuff as things are.
—WILLIAM JAMES, "DOES CONSCIOUSNESS
EXIST?" 1904

WE DO NOT KNOW WHERE OUR ABILITY to experience different tenses of time comes from and why we alone see time, past and present, in increments of weeks, years, centuries, and even more. Because we are the sole surviving branch of the genus *Homo*, it has proven hard to trace how our ability to engage in mental time travel arose and why. We routinely and consciously make physical records of our knowledge to distribute to distant times and faraway places inhabited by people we will never meet. Through culture, a collective form of memory, we create a shared view of the past that unites us into communities and allows large-scale cooperation among perfect strangers. All the evidence we have tells us that we alone have figured out how to do this. For all we know, this may be why we survived and our closest cousins, the Neanderthals and Denisovans, did not.

New techniques in DNA extraction and analysis allow us to compare our genome to those of our close cousins and see how their biological inheritance mirrors ours in most details.

Neanderthals were physically more powerful than our ancestors when we came out of Africa and moved into their territory. Neanderthals had been living in Eurasia for tens of thousands of years. They were better adapted to the cold, bigger in all dimensions including the brain. They may have shared some distinctively human behaviors that are now ours alone: controlling fire, crafting sophisticated tools, decorating their bodies, and even burying their dead. But Neanderthals did not band together in large groups to cooperate. They did not exchange goods across long distances, as early humans did. Nor did they create durable objects for the purposes of remembering and sharing information.

Our ancestors, on the other hand, communicated through language and gesture, song and dance. They led rich interior lives, creating ideas, images, and wholly imagined realities that shaped their interior worlds just as deftly as their tools made clothes to keep them warm and huts to shelter them. At some point in our past, we began to mark the passage of time. We started to see the relationship between action and reaction, cause and effect. We began to travel backward in time to understand causation, and forward in time to make predictions. We know that we did these things because we left material evidence of our thoughts. We began to *think with things*.

Over forty thousand years ago, *Homo sapiens* turned mind into matter by creating physical records of their mental lives. Strikingly realistic and profoundly sophisticated renderings of beasts and birds survive on the walls of caves in France, Spain, Italy, Germany, the steppes of Russia, and Indonesia. Objects that are less realistic but still recognizably human are found at related sites. These objects range from large caches of beads made from shell, bone, metals, and ivory to fertility fetishes and even musical instruments such as a flutelike instrument fashioned from a deer's bone. Items found at some sites bear traces of being manufactured far away, evidence that we were making and trading objects—along with women and livestock—at least seventy thousand years ago.

We can only conjecture what these images and objects meant and how they were used. But one thing is beyond doubt: whoever made and used them was human. They even signed some of their paintings, leaving handprints silhouetted in red and black tints. We recognize ourselves in the irrational yet compelling desire to breach the limits of time and space, to bear witness to our existence, and to speak to beings distant in time and space. We look at these images in the caves as a window into the past. What we find is a mirror.

Some scientists believe these paintings were made to invoke the spirits of hunting or fertility. Others argue that the caves were sites of shamanistic rituals. A third group thinks that the caves represent nothing more or less than Paleolithic graffiti. We cannot read the code in which they are written because we have no context for interpretation. In the absence of evidence that can be interpreted unambiguously, we continue to test our conjectures against any new evidence that might come in, sketchy as it may be. We demand that our knowledge be based squarely on the evidence at hand. This is a very modern mindset, this insistence on verifiable truths. Our ancestors were more direct in their explanation of how we came to be separate from other creatures: We asked one too many questions.

CREATION STORIES

How did memory grow from a feature common to invertebrates like snails and jellyfish to the elaborate cocoon of human culture? The unique mental capacities that make possible not just learning but also teaching and the wholesale accumulation of knowledge over millennia require self-awareness, symbolic thought, and language. The eerie certainty we have of existing as separate creatures in a world full of things that are not us, our ability to create abstract symbolic representations of our mental states and use language to communicate these interior states of mind—all this had to be in place before we could keep knowledge from dying with the person who possesses it.

Creation myths usually feature people who are dangerously curious. The generations of dangerously curious men we honor as the founders of modern science were raised on the biblical tale of origins, a calamitous account of curiosity and its consequences. Roger Bacon and Galileo Galilei, Francis Bacon and Isaac Newton, and Charles Darwin and Albert Einstein all grew up in a culture that instructed them that reverence must always trump knowledge. When Adam and Eve rashly found out for themselves what the fruit of the tree of knowledge of good and evil tasted like, the resulting breach of the natural order divided all time forever into Before and After. Having succumbed to curiosity, they sealed our fate.

In the Garden of Eden, there was neither toil nor trouble. All was provided for the well-being of God's creations. The Lord God instructed his special favorite, man, that "you may freely eat of every tree of the garden; but of the tree of the knowledge of good and evil you shall not eat, for in the day that you eat of it you shall die." Despite this prohibition and the explicit warning of deadly consequences, Adam and Eve had their apple. They chose know-ledge over reverence, and at that moment, history began. They were expelled from the Garden of Easy Living, permanently alien-ated from the natural, un-self-conscious state they were born into. The moral of the story is that in Paradise, there is no curiosity.

From that time on, we have been condemned to live by our wits. Culture is the collective wit by which we live. Having access to a collective memory that accumulates over time dramatically lowers the cost of our innate curiosity by providing a richer and more diverse set of possible answers to our endless questions. The story of Adam and Eve provides a wonderful thought experiment for us to learn exactly how all encompassing and cocoon-like culture is. In the starkest terms, Adam and Eve present a picture of humans who are all Nature and no nurture. Try to imagine their plight, suddenly cut off from the source of everything they need to live, thrown out on their own, having to fend for themselves with a newfound awareness of their own frailty and mortality. Without the vast legacy of know-how we take for granted as our birthright, they were forced to start from scratch and learn everything—how to find food, how to clothe and shelter themselves, how to bear children, how to raise them, how to die. Pain came as a surprise, and so did pleasure. They were like children—except they had no parents to turn to for explanations about what they were feeling.

Trying to imagine every aspect of life that Adam and Eve had to invent completely on their own makes clear how impossible it is to imagine being human in such circumstances. This premise has been the inspiration for enchanting fictions about the human condition, from Robinson Crusoe to Tarzan of the Apes. And in the midst of a new wave of apocalyptic thinking sweeping through culture at the beginning of the third millennium, some people are building libraries from which we could "restart civilization" in the event of the catastrophe they fear awaits us. These libraries are thought experiments too, ways of speculating about what we need to save and what we can afford to lose.

To be fair, Genesis conveys a contradictory message. Prior to having tasted of the fruit, Adam and Eve could not be said to be making a conscious choice to commit a sin. They were natural creatures, lacking any self-consciousness, ignorant of good and evil. They had no idea what death was. It did not exist in their experience, so when God said if they ate the fruit they would die, they may have had no idea what he was talking about. Their lack of experience is the very essence of their innocence. On the other hand, Adam and Eve were clearly endowed by God with their fateful curiosity—all that exists, exists by the grace and power of the Creator. So they may be guilty, but they are not ultimately responsible.

Whether Galileo or Darwin believed all or none of the story, it was the model of historical thinking that they learned as children and, as we will see, that tale had a powerful effect on how memory and knowledge were imagined in the West. Today the pursuit of curiosity for its own sake is hailed as the bedrock of our culture of knowledge. Through a curious interbreeding of biblical theology and Greek thought, the West gradually stopped seeing knowledge as a threat to reverence and instead began to cherish it as a form of reverence in itself. The story of how this change occurred unfolds over thousands of years and is full of twists, turns, and many dead ends. But it begins, humbly and inevitably, with accounting.

WHY THE ACCOUNTANTS INVENTED WRITING

Well before 3000 B.C., people were clustering into densely populated urban centers in Mesopotamia, today's Iraq (perhaps the historical site of the mythical Eden). One group, living in a proto-city we know as Sumer, began to write receipts when they were conducting small trades. These receipts were not made of paper, which had not been invented yet. Instead, people would recruit someone skilled in the art of inscription to incise certain symbols, understood by both parties to the trade, on a cylindrical clay token, thereby creating an object as witness and warrant of the transaction. These tokens are known today as cuneiforms, from the Latin for wedge (*cuneus*), the shape a scribe's stylus would make in the clay.

Over time, these tokens became larger, flatter tablets of either dried or fired clay. The wedges developed into a vocabulary of words, objects, names, numbers, and actions. While evolving over thousands of years in the Middle East long after the Sumerians passed into history, writing remained intrinsic to the process of accounting for goods and services and making public—publishing— that information. Recorded information lasts only as long as the medium on which it lives. The more durable and secure the carrier of the information, the more durable and secure the information itself. There may have been forms of inscription that preceded the cuneiforms. But if so, they were on more fragile media and have left no trace. Clay, on the other hand, stabilized that information literally for millennia.

The Sumerians are credited with inventing writing, and they no doubt deserve a special place of honor in our memory. But cuneiforms represent a more powerful innovation in the deep history of memory than a technical solution. This innovation made the invention of writing not only possible, but also almost inevitable.

It led to the creation of *objects as evidence*, capable of transcending the frailty of human memory and thwarting the temptation to shade the truth by holding people accountable. The cuneiforms are objective witnesses that cannot lie or forget. These truth tokens mitigated the risk of exchanging goods and services with strangers. They were warrants of trust that were hard to tamper with. No one could alter what was written in clay without making a mess of the tablet. The invention of physical objects that provided information—which if not fully unimpeachable was certainly far more reliable than human testimony—opened up a world of possibilities that lay beyond the walls of the city. But the need came first, then the invention.

The earliest surviving cuneiforms were created before 3300 B.C., and soon—in Sumerian if not Internet time—an information culture centered on these objects began to flourish. The oldest tablets comprise inventories of goods such as grains, oils, textiles, and livestock. But like any good invention, writing began to reveal its latent powers with use. People turned to writing for much more than accounting. Succeeding royal governments created the full array of documents familiar today—treaties, laws and decrees, and records of military prowess. All these were created with clear public purposes: to manage economic assets, secure control over a ruler's possessions, and extol his power.

THE INEVITABLE INVENTION OF INFRASTRUCTURE

The cuneiform, and writing in general, was invented to solve a practical problem: how to keep track of goods and services in an increasingly complicated society. But over time, the very success of writing created its own problems. The proliferation of tablets with valuable information led to the vexing questions of storage, security, preservation, and cataloging—an early instance of a Big Data management problem. When assembled en masse by central authorities, the tablets became a logistical challenge. Scribes encountered a predictable array of administrative questions: Where do we put all these tablets? How do we retrieve the ones we need? How do we know they are all safe and in order? Who is allowed to look at them?

For a while, each question presented itself as a discrete problem that was probably answered ad hoc by the scribes who were busy at their tasks. But it is impractical to keep inventing solutions each time problems arise. Over time, routines developed around storage and retrieval, as well as each aspect of the writing process: how to prepare the tablets, how to make the styli, which shapes indicated which meanings, the layout of each tablet, whether the script was read from the bottom to the top or vice versa, from the left to the right or vice versa. Once created, conventions about who could read them and under what circumstances had to be laid out. Storage systems had the unenviable mandate to protect the tablets from fracture, breakage, and theft; at the same time they had to be arranged for ease of use. And when several hundred cuneiforms grew into several thousand, the simple system for search and retrieval that had worked well enough at one order of magnitude had to scale up by several more. The scribes needed unilateral decisions about filing them according to the date of acquisition, or the date of creation, or the topic that they covered, and so forth.

The creation of *organized knowledge* was a slow, laborious process, born of the continuous iteration between problem and solution, ideas and technologies, one pushing the other and creating the familiar feedback loop of technical innovation. In each case, we see the same cycle of development, an ongoing dynamic of technical innovation in the service of practical problem solving, which in turn demands organizational and intellectual adaptations. Specially designed buildings staffed with specially trained experts arose organically to become core information infrastructure. The need to provide physical security, preservation, inventory control, cataloging, shelving schemes—these are all steps in the evolution of each recording medium that repeat themselves with every innovation, from clay tablets to silicon chips. As we strive to record more and more information in more compact ways, we inevitably create increased complexity, in the hopeful assumption that increased complexity is the same thing as progress. It seldom looks like that in the beginning.

CULTURE IS MORE EFFICIENT THAN BIOLOGY

As a mechanism of adaptation, culture is far more efficient than biology. Genetic material is more or less fixed at the time of conception. The genome does not acquire new information from an animal's experiences of life. Learning modifies the nervous system of an animal, but not its DNA. The offspring of a learned animal, bacterium, or fungus will be born ignorant of what its parent knew. It will have to acquire that knowledge from scratch. From the perspective of culture, evolution itself looks like a very dumb system—uninformed by intelligence, indifferent to learning, profoundly wasteful. Every step in evolution is taken by chance, the result of a random mutation in the gene. An animal's environment can shape how its genetic instruction book is translated into action and behavior. Some genes are turned on and others stay mute in response to stimuli from the world. But in contrast to the flexibility of human culture, genetic adaptability is stunningly slow.

This also explains why the human capacity to transmit learning across multiple generations through the creation and transmission of knowledge objects is utterly freakish in Nature. That we can do so is itself a genetically encoded endowment, which means it is freakish but still completely natural. As anthropologist David Bidney noted,

Man is by nature a cultural animal, since he is a self-cultivating, self-reflective, "self-conditioning" animal and attains to the full development of his natural potentialities, and exercises his distinctively human functions only in so far as he lives the cultural life. As contrasted with other animals whose range of development is biologically limited or circumscribed, man is largely a self-formed animal capable of the most diverse forms of activity.

Our evolutionary niche is to be a generalist, capable of adapting to many different ecosystems. We come into the world incomplete and need to be programmed by the specific culture we are raised in to reach maturity. Because we have settled in so many diverse environments, it is not surprising that cultural distinctions can be so dramatic. What can be a delicacy in one food culture can be taboo in another. What one culture considers beautiful—women with extravagantly long necks, tiny feet, or waspish waists—can be seen by outsiders as grotesque. Before globalization, there were thousands of ways of being human, each with its own language, dress, kinship systems, counting methods, and food ways.

We should not let the naturalness of human culture blind us to its dazzling power. Our ability to amass know-how over genera-tions turns out to be a biological adaptation that trumps physical strength, speed, and size. Unlike other animals, we do not adapt to different environments by becoming different species. Even though we have spread across the globe and exhibit many physical adapta-tions that set us apart one from another—straight black hair versus curly red, dark skin versus pale, tall narrow frame versus short and big boned—we call ourselves in all our variety *Homo sapiens*.

Instead we adapt to our different environments by developing different cultures, with distinct languages, cuisines, social struc-tures, economic and political organizations, and belief systems. We have developed radically different iterations of the "human" in our efforts to adapt to environments as diverse as Arctic tundra, high plains, rainforests, and coastal zones. We all grow up in a culture, and that culture—even the fact that we are cultural creatures—is largely invisible to us until we encounter someone, some thing, or some idea that comes from a different culture. It is telling that before globalization, when cultures could live in perfect isolation one from another, it was common for those who traveled across climate zones, mountains, or oceans to report that the human beings they met there were not actually fully human. As they moved across the globe, European explorers discovered populations that were so alien to them that they viewed them as inferior versions of humankind. It was just as common for those encountering the Europeans to think they were not human at all, but demons, or occasionally gods.

Because we are by nature culture-making creatures, distinctions we like to draw between what is natural and what is artificial or man-made are illusory at best. Culture mediates all of our experiences simply by providing each of us the basic template or mental model by which we interpret the world. Any feeling of there being a gap between humans and Nature is itself a by-product of culture. It is widespread in the developed countries of the West, but not in all parts of the world. We often blame the feeling of alienation from Nature on our technologies. But that feeling of separation existed well before the invention of computers, cars, and air-conditioning. Assigning blame to our tools and technologies is a by-product of secularization. Why we have that feeling is "explained" in Genesis as part of the human condition. But having rejected that explanation, we simply turned to technology as the culprit.

But there was nothing accidental or unnatural about the invention of writing in Mesopotamia. It had to have come naturally to us, because writing was invented multiple times and in multiple places. The Egyptians, Chinese, and Mesoamericans each developed writing systems of astounding complexity and ingenuity quite independently of each other. Cultures coming into contact with these writing civilizations adopted what they found and adapted the script for their own purposes. Extending the reach and longevity of knowledge became a distinct competitive advantage not only over animals, but also over rival *Homo sapiens*.

The strategic alliance between knowledge and power, record keeping and administering power intensified over the centuries of rising and falling empires in Mesopotamia. The Assyrian potentate Ashurbanipal, who ruled from 668 to ca. 627 B.C. and was by all accounts a learned man, amassed a library in his palace at Nineveh (across the Tigris River from present-day Mosul). Over thirty thousand tablets survive from the ruins of the royal library and are housed today in the British Museum in London. By then, the long life of recorded knowledge was already a given. People took for granted that artificial memory could respond to needs, desires, and ambitions that were equal in importance to survival as food and water. Among the many legal, administrative, and financial records of the Assyrian state, the royal library held tablets inscribed with medical cures, astrological and calendrical charts, magic spells, divinations, prayers, and poetry. Given the fact that these tablets outlived people for generations, they became the go-to medium with which to communicate with spirits from the past, record predictions of the future, and immortalize their own desire to cheat death. What is the earliest poem we have from the cuneiform era? The *Epic of Gilgamesh*, written in the second millennium B.C. It is the story of man seeking to live forever.

CULTURE SLOWS THINGS DOWN AND SPEEDS THINGS UP

Michel de Montaigne cautions us that nothing in life is fixed and forever. All is in constant flux. "We, and our judgment, and all mortal things go on flowing and rolling unceasingly. Thus nothing certain can be established about one thing by another, both the judging and the judged being in continual change and motion." But memory helps in the adjustment. Autobiographical memory gives us a sense of who we are and provides continuity as we age. We filter information all of our lives, and as we age our filtering process changes too because, as Montaigne said, we change and the world changes. As we move through infancy, childhood, adolescence, maturity, and on into old age, the task of memory also changes. In the beginning of our lives, acquiring knowledge about the world and about ourselves is our primary task. Later in life, when we know ourselves and our world better, we experience fewer novelties, and the ones that we encounter are very often assimilated into previous similar experiences. Memory begins to focus less on learning new things than on integrating all that we have experienced and known to provide a sense of continuity between past and present selves. We ask ourselves how our lives turned out the way they did and examine past experiences in the light of what we now know, like rereading a mystery after we know how it turns out in order to understand how the story unfolded and why we missed important clues the first time. This is memory's task of retrospection, to integrate the knowledge that we have, to impute a sense of cause and effect to the events in our lives, and to offer a sense of meaning.

Culture provides the large-scale framework for memory and meaning. It aids in the creation of new knowledge, but it also acts as a filter that over time determines what is of long-term value from the social perspective. It does so by being very conservative, retaining behaviors, practices, beliefs, values, and knowledge over long periods of time, be it the United States stubbornly holding on to inches and quarts when the world has moved on to centimeters and liters, or the unbroken stewardship of religious sites for millennia. Natural memory is designed to be labile, flexible, easily modified or written over to suit new environments. Artificial memory is designed to be stable, fixed and unchanging, slow, and resilient, freeing up mental space for individuals to learn new things. We are all born into a culture, specific to a time and place, that provides a wealth of ready-made knowledge and know-how for us to use in making our way in the world without delay. Culture is the set of given, ready-made elements that make up a living past, and that living past determines the basic parameters of lives and the choices available to us over time. It provides the baseline context of order and meaning in our private lives by directing our attention to some things and steering us away from others. In the nineteenth century, middle-class women in the Western world were encouraged find a meaningful and well-ordered life in the private sphere as wife, mother, or in Catholic countries, through a religious vocation. Contemporary middle-class women are raised with the expectation that they will seek a meaningful and orderly life in the public sphere as well, earning their living and aspiring to professional success alongside men.

The poet Czeslaw Milosz cherished his individual identity, but he knew that it emerged from the larger culture of the country and century in which he was born and over which he had no control.

"The clothes I wear, the technological conveniences I use, the verified and the unverified scientific hypotheses I have been taught, are not mine but my century's, and at most, it is only a tinge of the individual which slips into one or another set of given, ready-made elements." We may choose to dissent from our own culture, or make the choice that Milosz did, to move from his native country, Poland, to another world altogether—in his case Northern California—to escape a repressive political regime that threatened his very sense of autonomy. Given how much of our individuality emerges from our native culture, such a choice comes at a very high price.

Before the advent of the Internet, people had far less access to the diversity of human cultures that together make up the collective memory of humanity. On the one hand, from the perspective of our species, collective memory accelerates our adaptability to changing environments, lets us "pass Go" and slip the bonds of plodding evolutionary changes. It amplifies the human potential, both physical and mental. On the other hand, shared memory is the midwife of innovation and, paradoxically, accelerates the change in our environment. When we encounter a problem new to us, for example, we never have to start from scratch. We start from where the last people working on the problem (or a similar problem) left off. We recycle that solution, take it apart, tinker with it, and retool it to solve some novel problem or serve some new purpose. Without culture having captured and preserved that solution to begin with, we would have to start from scratch, just like the accursed Adam and Eve. After the Sumerians invented the cuneiform, no society having any contact with them, their successors, neighbors, or trading partners had to start writing from scratch.

Collective memory and the sheer power of knowledge accumulated over millennia both push us ahead and pull us from behind. Our knowledge allows us to make change readily, but in turn it forces us to adapt to ever-increasing innovations. Because of these dual forces acting on us, at moments of change we can find ourselves as a society behaving a bit like Doctor Dolittle's animal friend, the legendary pushmi-pullyu (push me–pull you), endowed with two heads—that of a gazelle at one end and of a unicorn at the other. The dear creature would often find himself processing the same information with two different minds, making two discrete decisions about what to do, and taking off in two different directions at the same time. If he tried to move in opposite directions, he struggled and stalled. And if he moved in two similar but still different directions, he would move along a path that was a little bit of both and end up where neither had expected to go. The faster he tried to move, the less predictable his destination became.

In the present day, with the double whammy of rapid changes in both the natural and social worlds, it can feel like our collective memory is a drag on our ability to adapt fast enough, binding us to a dying past when we are eager to move into the future. But the slower we move, the more control we gain over our final destination. In periods of great instability, the past becomes more useful as we increasingly tap into the strategic reserve of humanity's knowledge. Yet it is at moments like this when the past is most easily lost.

CHAPTER THREE

WHAT THE GREEKS THOUGHT: FROM ACCOUNTING TO AESTHETICS

JULIUS CAESAR IS FAMOUS for writing books and for burning books. The account of his conquests of Gaul and Britain is in print over two thousand years after he wrote it, translated into hundreds of languages, and still widely used as a primer for Latin students. His history of the civil war waged against former friend Pompey and foes in the Roman Senate is still a point of departure for anyone wanting a contemporary account of the Roman Empire's chaotic birth. His success as an author exemplifies the adage that history is written by the victors. They determine what the "past" looks like to succeeding generations.

In our day, Caesar is equally—and incorrectly—famous for having deprived us of a vast and literally irreplaceable legacy of classical literature. In 48 B.C., on a campaign to help Cleopatra in her battle to subdue her brother-husband-rival Ptolemy XIII, Caesar set fire to the ships in Alexandria's harbor to thwart an enemy incursion by sea. It is likely that sparks from the naval firestorm at sea were driven by the wind and landed on the roofs of the great Temple of the Muses (or Mouseion, whence our word "museum") that housed a collection of three hundred thousand scrolls—known today as the Library of Alexandria. These losses are a regrettable, if predictable, by-product of war. But Caesar is not to blame for the "Great Vanishing" of classical memory. Within a few years, Cleopatra's beloved Mark Antony restocked her library with two hundred thousand volumes pillaged from the great Library of Pergamum. At its peak in the following centuries, Alexandria's library held over half a million scrolls. The cultural amnesia induced by their complete loss was not Caesar's doing, but the work of many generations, Christian and Muslim, who felt no responsibility to care for pagan learning.

In his day, Julius Caesar was also renowned as a great orator, though neither his rhetoric nor his poems survive. Part of Caesar's success in defeating his rivals—military and political—was due to his oral powers of persuasion, his ability to paint incandescent pictures of the future greatness of Rome that inspired his fellow countrymen and senators and roused his troops with brilliant images of battlefield glory and the tangible rewards of victory. Thus by the first century B.C., the kinds of things that cultured people wanted to know and why had undergone a fundamental transformation, a transformation that we credit largely to the Greeks and that generations of Roman citizens and subjects adopted, adapted, and spread from Tunis to the Thames and beyond.

In the span of three thousand years, the inscription technologies pioneered by Sumerian scribes had evolved from pressing wedge-shaped styli into clay as tokens of business transactions to full-blown alphabetic writing on sheets of papyrus that look, for all intents and purposes, like the pages of a medieval manuscript. Three thousand years is a long time in human scale, roughly sixty-five or seventy generations if we generously estimate the life span of those reaching adulthood to be forty-five years. The increased urbanization, intensification of agricultural practices, and monumental building projects that produced Egyptian pyramids and the Acropolis of Athens testify how far humans left biological evolution behind them as the primary agent of change. If the Sumerian scribes could look down from some imaginary perch of space-time to behold their first century B.C. brethren, they would see two things quite beyond their imagination.

First was the remarkable new technology of writing that used inks to mark letters on sheets made from papyrus, a plant abundant in the marshlands of the delta of the lower Nile River but unknown in the drier climates of Mesopotamia. The sheets were made of papyrus stalks, pounded together into a fiber that was flexible, relatively strong, paper light, took inks well, was stable in dry climates, and could be rolled up into a tidy little scroll that was easier to handle and store than a clay tablet. They would also have been surprised that the symbols used to convey meaning had changed so dramatically over time, reduced essentially to letters made of a few strokes representing a sound and the letters together comprising an alphabet. The codes for writing had become more economical and made recombination more flexible.

On the other hand, they may have been dismayed to realize that the astounding efficiencies of ink-on-paper writing were bought at the price of durability. The clay tablets the Sumerians wrote on were subject to torching during wartime, just as scrolls burnt in the inferno that consumed the library of Alexandria. But fire does not destroy clay tablets. On the contrary, it preserves them. When exposed to heat in a process known to potters as firing, they became far more durable. An indication of just how stubbornly sturdy clay tablets are is that between five hundred thousand and two million clay tablets have been dug up over the past few centuries, many of them perfectly preserved. They may not be as lightweight and portable as scrolls, let alone books, CD-ROMs, or the silicon chips in our smartphones. But in terms of sheer durability, the technology for writing reached a peak five thousand years ago and has been going downhill ever since.

Our Sumerian scribes gazing down with curiosity at the Mediterranean world of the first century B.C. would also have been surprised by the sheer number of large collections and people who could read them. The literate were a bigger portion of the free population in Greece and Rome than they were likely to have been in Sumerian or Akkadian times. As experts on ancient slavery have pointed out, the freedom from labor enjoyed by the elites of Greece and Rome was bought and paid for by an increased dependence on enslaved labor. It is hard to avoid the conclusion that knowledge grew at the expense of human freedom, especially when we are used to thinking of the former being a precondition for the latter. But the equation was flipped only during the Enlightenment.

Libraries proliferated throughout the ancient world. Besides the collections in Alexandria, there were large libraries in Rhodes, Pergamum, Athens, and Rome, with many smaller collections in other population centers and private libraries in the homes of the cultured rich. But something else, perhaps less visible to the eye but ultimately of greater consequence, had shifted. By the fifth century B.C., the Greeks had embarked on a novel enterprise, the concerted cultivation of knowledge for its own sake. In doing so, they made three contributions to the expansion of human memory whose effects are still playing out today. The first is the creation of mnemonic or memory techniques that tap into a profound under-standing of how memory relies on emotion and spatialization, thereby predating contemporary neuroscience's findings by twenty-five hundred years. The second is the creation of libraries as centers of learning and scholarship, not primarily storage depots for administrative records. And third is recognition of the moral hazards of outsourcing the memory of a living, breathing, think-ing, and feeling person to any object whatsoever. By cultivating knowledge for its own sake, they raised the pursuit of beauty and harmony to a level as high as, or higher, than the pursuit of know-how to solve pragmatic problems. Knowledge was not cultivated solely for its instrumental value—its ability to effect change in the world. It acquired an aesthetic dimension, serving to give pleasure and meaning to individuals. And the Greeks insisted linking what we know with what we take responsibility for knowing as a moral matter, a question that has grown acute today, when scientific knowledge is used to create nuclear and biological weapons that can be dropped from robotic planes on civilians and military alike, and to extract energy and food for a ballooning human population that imperils the global environment.

MEMORY PALACES AND MNEMONICS

Cultures that do not develop writing have many ways other than the written word to enhance natural memory. Techniques to store vital information—about plants and animals, weather patterns and seasons, family and foe—are a central feature of all known cultures. They carry knowledge not only in songs, dance, and stories, but also in arrangements of rocks and patterns woven into a fabric. The Mano tribesmen of Liberia select stones to hold their memories. Incans knotted strings of llama silk into specific patterns governed by conventions of meaning, known as quipu, to convey information. These techniques require compressing sometimes lengthy pieces of information into a code. The contents of a message thus encoded could be shared with anyone with knowledge of the code, be it the single individual who created the memory stone or the group of Incans who communicated across their vast Andean empire in the language of knots.

The Greeks are credited with inventing a memory system, or mnemonic (so named after the Greek goddess of memory, Mnemosyne), that is still in use today. Along the way, they made two fundamental discoveries about how the brain forms memories through emotion and spatial thinking. Prizing the art of rhetoric as a civic virtue—democratic citizenship in action, as it were—the Greeks had to perform feats of memorization and recitation. So they built virtual libraries in their minds to store large components of complex knowledge for ease of recall in performance. These virtual libraries came to be known as memory palaces. Each orator was responsible for building his own personal mental memory palace and stocking it with information of his choosing—rhetoric to be delivered at a banquet or an ode to commemorate a victory. (The public sphere was for men only.) Once constructed, the memory palace enabled its architect to retrieve memories as easily as imagining a stroll through this imaginary edifice.

The legendary systematizer of memories and father of mnemonics was the lyric poet Simonides of Ceos (ca. 556–468 B.C.). He was not a man with an unusually good memory, but a man who discovered an unusually good technique for remembering things. Simonides was dining one day with a number of eminent folks when, midbanquet, he was called outside to meet two men. When he went out, he found no one. As he turned back, he saw the building he had just left collapse, crushing everybody inside. The bodies retrieved from the rubble were battered beyond recognition. Yet when called upon to identify the bodies, Simonides was able to recall who sat where as they dined. He identified people by the location of their seats around the table.

The first thing to notice is that his memory was *formed under extreme duress*, the trauma of having escaped a certain death. Emotion in memory formation and retention is primary. It lends brightness to the details of an event and cues the mind to the value of the memory's content. Emotion is embedded in the content of the event and is part and parcel of what is recalled when a memory is triggered. This affective or emotional nature of memory is responsible for the vividness of painful memories (a crippling vividness in the case of post-traumatic stress disorder). But it is equally the source of great pleasure when, catching the strains of an old song, you remember a certain blissful summer day at the beach when the radio played that tune over and over; or when the smell of caramelized apples takes you back to your grandmother's kitchen, warm, cozy, and pleasantly humid while outside darkness gathers at the windows and the winds rustle the last leaves on the big red maple by the garden path.

The second thing to notice is that Simonides recalled by *visualizing a spatial arrangement*. He summoned an image of where people sat around the banquet table in order to call their names to mind. This story, factual or not, makes absolute sense as neuroscience. Long before scientists studied the anatomy of memory, Greeks had figured out how emotion conveys the value of a memory and spatialization determines recall. Both declarative memory—facts, figures, names, dates, events—and spatial navigation are initially processed by the brain's hippocampus, a small neural structure in the brain. This is the part of the brain Alzheimer's disease attacks first, producing both memory loss and spatial disorientation among the afflicted. (In 2014, the Nobel Prize in Physiology or Medicine was awarded to three scientists who discovered how some aspects of this information mapping occur.) Without the careful placement of a perception into a spatial grid in the brain, the mind gets lost in the maze of memories crammed into our skulls. And without emotional encoding of value and vividness, that event and every scrap of information associated with it disappears into the anonymous oceans of daily data.

From this discovery of a striking visualization inflected by emotion, Simonides developed his technique. Generations of Greeks, Romans, and Europeans who revived this technique in the Renaissance used his system, based on the imaginary disposition of objects, standing in for ideas, carefully arrayed in an arbitrary space they imagined moving through to reach those ideas. Cicero, a devotee of the technique, wrote:

> [Simonides] inferred that persons desiring to train this faculty [of memory] must select places and form mental images of the things they wish to remember and store those images in the places, so that the order of the places will preserve the order of the things, and the image of the things will denote the things themselves, and we shall employ the places and images respectively as a wax writing-tablet and the letters written on it.

Objects were charged with symbolizing the content of a memory, and places were designated as navigation points along routes of retrieval. You could design a building full of long corridors punctuated by doors leading into rooms stocked with specific memories in the guise of designated objects. A raven could stand in for a vision you had in a dream. An armchair might recall the contents of a book you read. The large cannon in the middle of the room might summon up a visit you made to Gettysburg one summer and this puts you in mind of the garden in Delaware you saw on the same trip, the garden you were strolling in when you first heard the story about how your parents met.

Objects alone cannot do the trick. Instead, we use a compression algorithm native to symbolic thought, the use of the concrete as a part to stand in for the whole. In poetics, this part-for-the-whole substitution is termed "synecdoche," and it proliferates in our language because it is so vivid and efficient. We call the executive branch of the government the White House and the legislative branch Capitol Hill, for example, and government business passing between them is said to go down Pennsylvania Avenue, the name of the boulevard that runs between them. The art of memory simply leverages this mental shortcut.

Recall depends on connections among memories dispersed across far-flung networks of stored information. When we search for a specific memory, we are like fishermen casting nets overboard into the fathomless seas. What we haul up is not any one name or place or event, but a vast web of assorted yet tightly associated facts and figures, contents from the depths of our past jumbled up with the flotsam and jetsam of yesterday's news. We do not notice all this detritus because we are looking for one thing and ignoring the rest. Recall aided by spatial memory is common, for example, when we try to remember personal names. When we encounter a familiar-looking woman but cannot remember her name, we grope for where we last saw her to recall her name. In a split-second search and retrieval, our mind may fill up with all kinds of associations, but we are not conscious of that because we are focused on spotting just that one name.

For mnemonics to work best, the location in which we place memories should be arbitrary, neither logical nor natural. It could be a landscape—a forest, a garden, a mountain path—anything with strong features easily remarked and remembered. But most people practicing the art of memory choose to construct imaginary buildings. This gives them more control than having to adapt their mental landscapes to the real world. During the Renaissance, when people were expected to be eloquent without resort to teleprompters and even cheap notepaper was nonexistent, hands were used as sites for memorization. The palm was the handiest possible portable mnemonic device. For reasons we do not fully understand, memory can be reinforced and amplified by using physical objects, whether it is a memory stone, a series of knots tied on fingers or into elaborate quipu, or merely an extension of the body itself.

The power of spatial context to aid memory retrieval is the principle behind many residential-care regimes for Alzheimer's patients. The staff essentially re-create the context of a person's life by re-creating a room in which they lived. Familiar surroundings cue the patients that they are somewhere safe. The effect is to orient them, calming them by obviating the need to learn an unfamiliar landscape.

The routine of walking through a real space also works to stimulate recollection. Locomotion gets more than just the body moving. It is common for mathematicians, composers, writers, scientists, and anyone else engaged in mental exertion to take a walk when they are stuck, as if moving can literally jog free something that will not budge in their brain. Beethoven, Dickens, and Kierkegaard were all devotees of the long afternoon stroll. We do not know why, let alone how, moderate physical movement stirs up the archives of the mind along with the circulation of the blood. We do know that when researchers long used to browsing library shelves to find something (even if they do not know what it is) complain about missing physical browsing in the online environment, they are doing more than lamenting the loss of a search technique they feel comfortable with. We may refer to the Internet as cyberspace, but its lack of material substance has distinct disadvantages when it comes to finding our way in its dense forests of data. We understand so very little about how real physical space affects memory and vice versa. We know Alzheimer's patients lose their ability to navigate space and even orient themselves physically in the present, for example, unable to tell where they are because they cannot remember where they came from. And we know the hippocampus maps location cell by individual cell. But how that map is processed into memory remains obscure.

Fortunately, maybe even predictably, we have rediscovered the importance of geography and the art of mapping in the digital age. Virtual representation of space is necessary in the absence of real space because the brain spatializes what it perceives. It is not accidental that many Internet search engines use some form of map to display results. Digital search techniques are good at identity matching, whether we are searching databases for genetic or fingerprint matches, or looking for hypertext links that work through verbal matching and association. But the sense of meaning only arises from the context of what we perceive. Interactive timelines, maps, charts, and infographics are common now in presentations and online newspapers. They allow people to grasp the import of information quickly by arranging data schematically in a variety of contexts to reveal relationships. Context is spatial. Simonides knew that.

LIBRARIES AS TEMPLES OF LEARNING

The volume of information crossing someone's horizon is always limited by what is physically accessible. Until roughly 2000 A.D., if someone wanted access to information, they had to go to where the books and journals, maps and manuscripts were—the library. What the Greeks started and the Romans elaborated was the transformation of a library from a depot of administrative records to a workshop for knowledge creation.

The library at Alexandria was built on a strip of land where the Nile delta meets the Mediterranean. The city served as the gateway to the interior of the country and Egypt's portal to the world. What we call the library was a collection of manuscripts housed in a separate building within a temple complex dedicated to the muses. The library's birth and death are shrouded in legends, but we think the library was created sometime in the fourth or third century B.C. and died a slow death from repeated armed assaults, collateral damage of war, followed by studied neglect over the course of many centuries as it was dispersed or destroyed and eventually disappeared. The library has no death date. It did not really die. It faded away.

In its heyday, it served as a gathering point for copying documents from around the Mediterranean. Egyptian rulers would "borrow" manuscripts from vessels that came into port and copy them for the collections. (They often kept the originals and returned copies to their owners.) At its peak in the early centuries of the Roman Empire, the collection may have been as large as seven hundred thousand scrolls comprising tens of thousands of individual works, though an unknown number of those were likely to be variants or duplicates.

It was a collection that supported scholarship, embedded in a temple to learning. The collections were copied and managed by experts, studied and edited by other experts. The scholars were paid for their intellectual labor, the manuscripts collected and cared for as their tools of production. The legend of Alexandria is populated with great minds affiliated with the Mouseion over centuries, advancing scholarship in a culture that cultivated knowledge for its own sake—among them Euclid, Archimedes, Eratosthenes, Galen, and Hypatia. For all these reasons, the legendary library glows with the charisma of an academic utopia, a professor's dream of the Garden of Eden, where all bodily and intellectual needs are provided for—particularly as the scholars, while they may have taken on private students, did not have teaching obligations. The legend must remain a legend, for there is no useful archaeological evidence of the library to shed light on its long and turbulent history.

Good work requires good order. Before scholars could begin their work, the librarians had to have finished theirs. Their task was twofold: to create and maintain an intellectual organization so it would be possible to find things, and to provide physical stewardship, to keep the scrolls secure and usable over time. There was probably no bright line separating librarian from scholar. (The professional distinction between the two is a very recent development, dating only to the mid-twentieth century.) The collections were spatially arranged to form an intellectual order and hierarchy. Individual rooms were set aside for specific topics and, within that, arranged alphabetically by name of author. It does not matter how comprehensive and well tended a collection may be. If an item cannot be located on demand because it is out of order, misplaced, or incorrectly cataloged, it effectively does not exist.

Greater efficiencies in the production of scrolls over time meant the proliferation of physical objects that were valuable, fragile, and, once rolled up, identical in appearance. Thus rolled, scrolls can be safely stored and stacked one atop the other. But the problem is that when stored efficiently they all look the same. So scribes started attaching tags to the ends of the rollers for quick identification. So far so good. When you were done with one, you put it back on top of the stack it came from. But what if you wanted to look at the scroll at the bottom of the pile? You would have to carefully adjust everything on top to free the one on the bottom. Manual labor was in abundant supply in the ancient world—slavery ubiquitous and the employment of children routine. The librarians of Alexandria could afford to solve the scroll-management problem by throwing a lot of cheap labor at it. But a better solution was a technically advanced format—the codex. This is the book form we know today, sheets of paper (or papyrus) cut into uniform size and bound between covers. The codex is much easier to search than the scroll, more portable, and because it has a spine, it is much easier to stand up on a shelf and affix the title to the spine for easy identification. This new book technology was more efficient both in keeping good intellectual order, and in maximizing use of space on the shelf.

Until the present age, managing physical objects was the only way we managed knowledge. The order of knowledge mirrored the order of things. In digital archives, there are no objects, only bits. They are stored randomly and assembled on the fly when they are called up to the screen. There is no deliberate spatial arrangement of bits on a chip. But then again, it is only a machine that searches the archives. Now we throw a lot of machine labor at the problem of storage and retrieval.

The Library of Alexandria has been lionized in recent history as having had comprehensive or universal ambitions, collecting knowledge on all subjects comprehensively, in multiple languages, and not according simply to one ideology, religion, or other tribal sense of inclusion/exclusion. It was an imperial library in what was probably the most cosmopolitan city of the ancient world—Greeks,

Romans, Egyptians, Jews, Armenians, Persians, and more all living cheek by jowl. As depositories of human memory, libraries became the symbol of man's attempts to master the world through the gathering of all knowledge. No library was quite as ambitious as the one in Alexandria. It is the essential model for the library in the digital age.

THE MORALITY OF MEMORY VS. THE EFFICIENCY OF WRITING

Greeks embraced the necessity of writing things down so that knowledge may endure for generations, but not without making a philosophical issue out of it. In the fifth century B.C., according to the written testimony of Plato, Socrates warned that the invention of writing would lead to ignorance and, ultimately, the death of memory.

> For this invention will produce forgetfulness in the minds of those who learn to use it, because they will not practice their memory. Their trust in writing, produced by external characters which are no part of themselves, will discourage the use of their own memory within them. You have invented an elixir not of memory, but of reminding; and you offer your pupils the appearance of wisdom, not true wisdom, for they will read many things without instruction and will therefore seem to know many things, when they are for the most part ignorant and hard to get along with, since they are not wise, but only appear wise.

Once knowledge is transferred to a piece of paper, then it essentially leaves us and with that, Socrates argues, we no longer feel responsible for remembering it.

Writing has its uses, of course. Even Socrates may have taken comfort in the fact that Plato wrote down what he said, and so Socrates had a shot at immortality. The works of other Greeks—

the missing plays of Aeschylus, the lost verses of Sappho, the unknown writings of Democritus and Epicurus—are gone. Did we lose them *because* they were written down?

The key question Socrates poses to us today is about responsibility. Who is responsible for keeping knowledge alive? We keep our contacts, full of other people's personal data, on our smartphones which we then back up to the cloud, trusting that the information is secure and will be there when we need it. But where does that trust come from? For Socrates, remembering is a moral action, touching the very substance of our being. Remembering lots of phone numbers may not be the best use of our memory, but it is sobering to realize that even something as simple as storing other people's phone numbers and addresses with commercial services has potential moral implications. How do we know that the data we store in the cloud are safe from hackers, thieves, and surveillance agencies? Understanding the full risks and benefits of relying on a burgeoning array of memory devices is difficult in the digital realm. Change happens so rapidly that we seldom have the time to slow down and learn about the true costs of new services that free up so much time and mental space. Instead, we learn through trial and error.

When we commit something to memory, we absorb it, metabolize it, incorporate it into our mental model of the world. Memory was the foundation of the feats of rhetorical performance so treasured by the Greeks. But rote memorization and repetition were not what they had in mind. The art of memory was taught as a species of performance, something done in real time according to well-practiced routines. The very foundation of memory itself was understood to be emergent and performative—not fixed and forever, but coming into being under specific circumstances.

The goddess of memory, Mnemosyne, was daughter of Gaia (the Earth) and Uranus (the sky). Through a union with Zeus, she gave birth to the nine muses—of epic poetry, lyric poetry, song, dance, sacred lyrics, comedy, tragedy, history, and astronomy. These daughter divinities of the arts and sciences were honored at cult sites such as the one at Alexandria that housed the famous

library, places where knowledge was created, publicly demon-
strated, and thereby kept alive. Making knowledge public was
not accomplished only by distributing written copies of a text.
Publication was a bit more like the broadcast medium of the web:
Things were in constant circulation. If not, they fell out of common
discourse. Knowledge came alive through the spirit, the breath,
the inspiration that a muse breathed into a person (*inspirare* being
Latin for "breathe into").

The scholars and artists who brought knowledge into the light
of day relied on deep stores of ideas, words, melodies, images, and
equations committed to memory. The muses were invoked to
inspire or prompt the performers. These performers worked from
memory but were not slaves to literal memorization. Their versions
of a tale or poem would be individual. Like jazz artists, they could
move through a script they carried in their heads, fill in blank
spots, and riff on stock metaphors or passages, all the time bringing
something specific, personal, and unique to the audience. The
classical art of memory, practiced well into the age of literacy, was
an art of moral development through proximity to and incorpora-
tion of inspired minds.

Today's arguments that reading on-screen, as opposed to from
the pages of a book, will deaden our empathy use the same logic as
Socrates did. Both place high value in reliving someone's thoughts
and words, either through recitation or reading, as vital to the
development of man's higher nature. People are worried that
without the printed page, we will lack tools to hone our empathy
and develop the mental habits necessary for living on a crowded
planet that will increasingly place calls on our compassion. But
they are conflating medium and message, just as Socrates did. It is
ironic that people now lace themselves up in Socrates's sandals,
ruing the loss of print culture as a backward step in our moral
development.

While Socrates warned against the loss of wisdom by outsourcing memory to papyrus (or computer chip), his prediction
that external memory systems would hurt us as a species completely
missed the mark. If we had not turned mind into matter, our

biological memory would have been stuck forever in the present, small of scale and leaving little behind when we die. But once committed to physical objects, knowledge was not only shared among individuals. It was also distributed—collected, copied, and traded, sometimes lent, often stolen. Distributing knowledge over generations and across continents is the closest we will ever come to creating a natural resource that cannot be exhausted and whose value actually increases with use.

That said, Socrates's misgivings were not entirely wrong. Expanding the scope of knowledge above and beyond a certain scale makes it impossible to achieve the single thing he thought mattered in life: to know thyself. We can expand the volume and reach of knowledge faster than our ability to know what we know. The real moral hazard of outsourcing is that we outrun our ability to predict the consequences of our actions and refuse to take responsibility for how knowledge is used. It is an old fear that haunts all cultures of knowledge and keeps the tales of Faust and Frankenstein ever alive. It preoccupied the scientists working at Los Alamos on the atomic bomb. And it troubles almost all of us as we read accounts of genetically modified foods, computers that learn, and robots that speak in human voices.

KNOWLEDGE GOES THE WAY OF ALL FLESH

Books, scrolls, and cuneiforms are painfully vulnerable. The destruction of knowledge happens with frightening ease. Liquids can damage paper, clay tablets can shatter, damp climates can infect anything with mold. Careless handling of candles and hot wax can set manuscripts alight. Armed conflict puts knowledge at risk too, for libraries are most commonly located in urban centers or attached to seats of power that are the prize sought by invading armies. What survives war can die from inattention and neglect. The library at Alexandria ultimately collapsed not directly from wars. It died because people stopped valuing its contents. The ideological allegiances of both Christian and Muslim populations blinded them to the value of pagan learning. They had no need for the books that held the memory of the classical world in reserve for the benefit of future generations. Even if the Christian and Muslim leaders had not condemned pagan thought for impiety, knowledge for its own sake became unaffordable. Severe economic pressures and raging plagues that swept through lands once united under the Pax Romana meant people battled to live another day. Parochialism and short-term thinking tightened their grip on the public mind.

What we see clearly in the wake of classical literature's Great Vanishing is that the collective memory of humanity is dependent on two things: a durable medium on which to record an image, text, map, or musical score; and an institution, some organization that takes responsibility for the care and handling of the collection for generations into the future. The temple to the daughters of Mnemosyne in Alexandria was dedicated to the making and sharing of knowledge. When the knowledge trade itself atrophied, the museum and library no longer returned value to its legatees. They stopped investing in it.

Fundamental to today's anxiety about the future of memory is the lurking awareness that our recording medium of choice, the silicon chip, is vulnerable to decay, accidental deletion, and overwriting. And we know there are few institutions—if any— that have the scale and capacity to keep our analog legacy of knowledge intact at the same time they scale up to acquire the digital. This is a reasonable anxiety. Without preservation, there is no access.

Like preservation and access, the fate of media and the fate of institutions are so closely intertwined that the story of one can only be told as the story of the other. With every innovation in information technology that produces greater efficiency by further compressing data, librarians and archivists begin a frantic race against time to save the new media, inevitably more ephemeral. The next inflection point in the history of memory, the Renaissance and the invention of printing, comes one thousand years after the dissolution of the Roman Empire. It is quite wrong to refer to the period after that collapse as the Dark Ages. While Western

Europe fell victim to a radical case of cultural amnesia, in both Constantinople and centers of power and enlightenment in the Islamic world, much of the legacy of classical Greece and Rome was kept intact.

Therefore, when the self-styled humanists of the fifteenth and sixteenth centuries went in search of the classical legacy, their excavations of ancient sites and texts bore remarkable fruit. The revived and downright rambunctious spirit of curiosity that marked the humanists spawned its own information inflation as they uncovered text after text and created a voluminous body of critical writings to argue about what they found. The invention of movable type by a goldsmith in a town north of the Alps in the 1450s allowed multiple printings of a text to replace the one-offs of the scriptorium. Johannes Gutenberg and his shop inadvertently created in the printing press an accelerant that turned the modest flames of scholarly passion into the great intellectual conflagration of the Renaissance that grew quickly and spread the spirit of rebirth and reformation.

CHAPTER FOUR

WHERE DEAD PEOPLE TALK

*I seek in books only to give myself pleasure by honest amusement; or if
I study, I seek only the learning that treats of the knowledge of myself
and instructs me in how to die well and live well.*

—MICHEL DE MONTAIGNE, "OF BOOKS"

MICHEL DE MONTAIGNE (1533–1592) grew up with books, and his attachment to the pleasures of reading began early in life. The landscape of memory had only decades before been altered beyond recognition by two tectonic forces. The first was the excavation of long-buried literature and art from the ancient world, begun by an avant-garde of clerics and literary types in the fourteenth century. The Renaissance was midwife to the rebirth of a civilization that survived in complicated fragments and bereft of a broader context that would make it comprehensible. The second force for change was the technology of movable type. It was created in Mainz, in present-day Germany, on the periphery of the ancient world. But the Germans more than made up for their provincial status by the mighty force of their artisans' brilliant craftsmanship and dedication to technical perfection.

Printing presses arose to meet an insatiable demand for reading material. They in turn inflamed the curiosity of a new literati who wanted more books, produced faster and cheaper. Book production soon reached a scale that was unprecedented. By 1500, a mere four decades after printing presses began operations, between 150 and 200 million books flooded the circulation system of European culture. Both the quantity and the quality of what was available to read had taken a dramatic and surprising turn.

The result was that in the 1530s, Montaigne learned to read as a "print native," a creature thriving in the new ecology of memory and learning. He grew up in the midst of the first well-documented information inflation (it was literally self-documenting) and, through a series of bestselling books of essays published over the course of his life, he was both a mirror of his times and helped to shape them. The great inflation of information, carried far and wide in books and broadsides (the precursor of today's newspapers), created what seemed to many contemporaries, Montaigne among them, a promiscuous intercourse of ideas that was both empowering and bewildering.

Although the bookmakers were still learning to take advantage of the new technology's navigation tools (including such novelties as title pages, tables of contents, and consistent pagination), books were no longer designed as knockoffs of handwritten manuscripts. The first printed volumes had been skilled imitations of the handwritten codex. Today, in the Great Hall of the Library of Congress, two Bibles face each other in imposing exhibit cases the size of small cars, each produced within years of each other, and both masterpieces of their form. The Giant Bible of Mainz was produced the old-fashioned way by hand in 1452 to 1453. Across from it is a rare vellum copy of Gutenberg's first printed work, the Bible, probably produced in the same town about the same time and executed in a perfect imitation of the Gothic script used in scriptoria. With one glance you can see how perfectly Gutenberg's shop was able to create by machine a work that could pass for the unique product of a first-class scribal hand.

Montaigne seized the possibilities of the print form—which did not seem novel to him, after all—to play around with a genre of expression that found a ready audience, one that is digressive, almost improvisatory. He published and republished volumes of short prose pieces—three books totaling 107 individual entries, ranging in length from under a page ("On the frugality of the ancients") to well over one hundred pages ("Apology for Raymond Sebond"). He called his prose pieces *essais*, meaning attempts, tests, or perhaps experiments.

"What do I know?" he asks himself. Writing and revising his essays was his way of sifting the world for answers. Although in many ways an exceptional man—born into great wealth, highly literate in an age when less than ten percent of the population could read and write and was essentially all male—and quick to let his readers in on his peculiarities and unusual tastes, Montaigne saw in himself what was universal in the human condition. He believed that a full and honest account of himself would reveal something about ourselves we could not otherwise see, as if by holding a mirror up to himself, we could better see ourselves. This must be true, because Montaigne has never gone out of fashion or out of print. Each age seems to find something in his writings that appears to speak directly to them. For us, living through our own information technology upheaval and the religious and ideological wars waged online and in the flesh, Montaigne can exemplify the struggle we all face to identify what is true and what merely passes for truth because it is what we want to hear. He instructs us on how to filter out the distracting, distressing, irrelevant, or dishonest but self-gratifying. And he shows us that sometimes the best way to understand ourselves is to reveal ourselves to others, a lesson taken to heart today by a new generation of memoirists.

The essays began as a form of engaged reading, a conversation with friends who lived on in the books they wrote. While he was an eager reader as a child, spending time in the company of his favorite authors became a serious preoccupation for Montaigne only after his closest friend, Etienne de la Boétie, died and left him his own book collection. The year was 1563, and Montaigne was thirty years old. He was a jurist at the time, serving in the court of the provincial capital of Bordeaux. Eight years later he retreated from public life to his ancestral home in the wine-producing countryside, thirty miles east of Bordeaux. He declared himself officially retired. Here he spent his time reading, writing, and learning "how to die well and to live well." Montaigne reached a dead end when he felt he could not go on because his world had failed. A host of beloveds—friend, father, brother, children—were all dead and now his country was coming apart in civil war. He retreated in mourning and prepared to meet death nobly.

He wrote up an announcement in Latin and affixed it to the wall near his library.

> In the year of Christ 1571, at the age of thirty-eight, on the last day of February, his birthday, Michel de Montaigne, long weary of the servitude of the court and of public employments, while still entire, retired to the bosom of the learned virgins [the muses], where in calm and freedom from all cares he will spend what little remains of his life, now more than half run out. If the fates permit, he will complete this abode, this sweet ancestral retreat; and he has consecrated to it his freedom, tranquility, and leisure.

He spent most of his days at home in his chateau, secure in a corner tower he claimed for himself. His library was on the third floor, and in it he arranged his collection of close to one thousand volumes so that they were within easy reach. "The shape of my library is round, the only flat side being the part needed for my table and chair; and curving round me it presents at a glance all my books, arranged in five rows of shelves on all sides. It offers free and rich views in three directions, and sixteen paces of free space in diameter."

Montaigne was only five years older than his dear friend Boétie was when he died. But he felt old. Montaigne had suffered one loss after another. Boétie died after a sudden onset of dysentery in 1563. His beloved father died in 1568 from kidney stones after terrible suffering. His younger brother died a year later after a freak accident with a tennis ball. The next year his first child was

born and died within three months. And the losses did not stop
there: Out of six born, only one of his children lived to adulthood.
He himself had a close encounter with death sometime in 1569 or
1570, when his horse was accidentally rammed by another rider.
Montaigne was thrown to the ground and lay unconscious for
a while. He eventually came to, only to suffer convulsions. He
thought he was dying.

But he survived this, just as he had survived the deaths of those
closest to him. By his midforties, he experienced his first severely
painful attack of kidney stones, the disease that had crippled and
eventually killed his father. From then on, the fear of his imminent
death never left him. The pain from a kidney stone attack can be
so intense that people pass out from it, as Montaigne repeatedly
witnessed his father do. Until the stones pass, the patient is at risk
of inflammation and infection. Ultimately death can ensue. Today
there are safe techniques for breaking up the stones to allow easier
passage, and in the event an infection sets in, it can be treated with
antibiotics. There was no effective treatment in Montaigne's time,
although he tried every one he heard rumor of.

In the meantime, he would suffer excruciating pain whenever
they acted up. Worse, he realized, was that the memory of the
attacks played a cruel trick on him. When he was not suffering
attacks, he was suffering from the anxiety of anticipating them. He
confessed that "the thing I fear the most is fear." The paralysis of
anxiety, familiar to all who suffer anxiety disorders, was in its own
way more unbearable to Montaigne than physical suffering. He
finally set his mind to conquering that fear and developed what we
would call today a cognitive therapy that proved quite effective.
He consciously rewrote his memories of the event to focus not on
the pain of the attacks, but the intense pleasure felt upon their
relief. Creating an aide-mémoire, he writes down on paper ("for
lack of a natural memory") each step of an attack and how it passes.
When he feels one come on, he reviews the notes from previous
attacks and assures himself that this too will end. Familiarity with
past experience gives him hope. Toward the end of his life, when
he has faced many such episodes, he comes to the conviction that

"nature has lent us pain for the honor and service of pleasure and painlessness." Thus his experience, tempered by reason, taught him to live.

MONTAIGNE ASKS HIS FRIENDS FOR HELP

Montaigne was intimately familiar with the information overload produced by printed books. But as a print native, he had already devised a way to impose a filter between himself and information inundation. Like many of us today facing too much information, he turned to his friends for advice about what to heed and what to ignore. As a reader, he had developed an extended network of similarly bookish friends ever since he labored over his Latin lessons as a young boy. By the time he reached maturity, though, those he loved most dearly were no longer among the living. This was a sad regret, perhaps, but no cause for despair. He still had a library full of trusted friends who showed him by example how to die well and to live well, such friends as Lucretius, Cicero, and Socrates. When he lost Boétie to an early death, Montaigne turned instead to his library, the land of memory where dead people talk.

His initial essays are wonderfully death obsessed, with such cheery titles as "Of sadness," "Of fear," and "That to philosophize is to learn to die" (the latter a quote from Cicero). Each was rich with long citations from Stoic and Epicurean philosophers who had confronted death without the consolation of the Christian revelation. Reading and writing, which for Montaigne were almost inseparable, were themselves acts of memory, sustaining a conversation he had begun with Boétie and that he was simply forced to continue in a different mode after his friend died.

When he decided to publish the essays, he claimed to be writing for his own posterity. In the foreword to his first volume, published when he was forty-seven, Montaigne says that he has written his essays for "my relatives and friends, so that when they have lost me (as soon they must), they may recover here some features of my habits and temperament, and by this means keep the knowledge that they have had of me more complete and alive." It was a bold enterprise, making oneself the entire subject of a book. It seems quite modern to us—thoroughly uncontroversial, even tiresome in its concerted egocentricity. But that is because we are his descendants.

Montaigne may not have been the first writer to preface his book with disingenuous disclaimers about his humble intentions, and he was certainly not the last. But over the course of decades of publishing, editing, and republishing his essays, he continues to explain to the reader what he intends to do and how we, as readers, should understand what he is doing. Montaigne struggles to reframe his enterprise and reinterpret himself as he matures and records changes in himself. Though his early essays are a bit self-conscious, timid, even awkward by his own standards, we get a sense of how difficult it is to work in a genre that you invent as you go along. The processes of reading, thinking, writing were all forced through the crucible of this new, ungainly form he labored in—and his writing can seem quite labored, even accounting for the stilted style of the antique language.

WHEN THE WORLD FAILS

He tested his experience and reason for soundness by setting them alongside those of classical philosophers. This testing of self against history's exemplars is another way to understand what Montaigne meant by the word "essai," derived from the same root as "assay," or examine by trial or experiment. Paradoxically, even though his idols of wisdom were pagan, Montaigne felt safer citing them as experts than his contemporaries. The Renaissance had not only recovered what remained of classical literature, it resurrected these texts and gave them legitimacy as a body of historical knowledge—and just in time. Classical authors remained marvelously agnostic during the religious wars. Exemplars of pre-Christian mentalities, they were immune to the contagion of ideological warfare, standing outside, if not actually above, the battles waged among competing Christian confessions. Montaigne lived and wrote in exceedingly dangerous times, the early decades of the Reformation, when Europe was tearing itself apart, country by country, region by region, family by family over life-and-death-and-afterlife issues—the nature of grace, free will, the sacraments, even the divinity of Jesus. Civil war between the Catholic loyalists and rebellious Protestants convulsed France in the 1550s and continued on and off for decades. (The wars did not stop until a few years after Montaigne died.)

The fact that his family embraced both Catholics and Protestants and he was a committed moderate in his Catholic confession made him a target of special vitriolic attention from both sides. The middle ground was the loneliest ground to defend. He remained a loyal Catholic and kept peace with siblings who were Protestant. But as time wore on, the sectarian conflict acquired its own involuted logic of blood and vengeance, sweeping up all charity and compassion into a vortex of violence. Several times he was forced to evacuate his home and take his family on the road when his own neighborhood was overrun by warring factions, accompanied, as always, by the dark shadow of the plague.

There was an all-out crisis of authority throughout Western Christendom as one institution after another began to fail. In England, Henry VIII declared his independence from papal authority, confiscated church property, and shut down monasteries. In France, the crown passed among contentious factions professing opposing religious allegiances with alarming frequency. Assassination became a way of settling political differences. At one point Montaigne himself was thrown into the Bastille by the enemies of his friends.

No institution failed more spectacularly than the papacy, in its day the sole global power of the Western world, basing its authority on direct descent from the apostle Peter. It had never recovered politically from a crisis in the fourteenth century, when seven legitimately anointed popes reigned in succession from Avignon in France, not Rome, and several other individuals laid claim to the Holy See in Rome. But the papacy's religious authority and reputation remained fast among the faithful until they encountered detailed (often printed) reports of moral corruption in the papal court, frequently accompanied by salacious drawings that grabbed the eyeballs of literate and illiterate alike and that, like porn throughout the ages, traveled faster than the speed of light and farther than well-sourced and reliable reporting. People became disillusioned with the clergy at all levels. But they did not become less religious. On the contrary, it was an inflammation of religious passion that led to the reformation, rather than the dying off, of Christian faith.

The world of unified Christendom failed for many reasons, but technology played its part. The presses both facilitated the spread of antipapal propaganda and spread the word of the Gospels far and wide through multiple vernacular translations. On the one hand, printing helped the pope sell more indulgences (which awarded the purchaser a quicker passage to heaven after death). On the other hand, the presses produced a big run of Martin Luther's "Ninety-Five Theses on the Power and Efficacy of Indulgences," excoriating indulgences and all manner of papal venality. The presses had destroyed the possibility of monopolizing channels of communication. It took a few generations before people and parties became adept at controlling the presses so they could control the message. In the meantime, there was a scramble among publishing entrepreneurs to gain market share for their books, not unlike the jostling we see today for control of search, social media, and apps for mobile devices. Printers pioneered consumer-friendly features such as the small-format volumes (octavo size, the equivalent of today's paperback books), newfangled fonts designed for easier reading, and the liberal use of commas and semicolons to ease the speed of comprehension.

In reaction to the challenges to papal authority from Luther and the constellations of religious dissidents spun off from the implosion of doctrinal orthodoxy, the church convened the Council of Trent over the course of several years (1545–47, 1551–52, 1559–63) to plot a counterattack. Cornered and feeling vulnerable, the councilors began deciding matters religious and otherwise that had been left open, ambiguous, and adaptable. Now, wherever a good

Catholic looked, there was orthodoxy. The church even compiled a list of tainted titles, the Index of Prohibited Books, forbidden to the faithful. In 1676, Montaigne's essays were added to the dishonor roll, no doubt ensuring a steady readership over the centuries to come. The world of the all-merciful God, colored in subtle shades of gray, was becoming starkly black and white. While introspection for the sake of taking account of one's soul was encouraged by the church, trying to understand oneself as a person, a singular and unique individual with a worth and dignity distinct from one's soul—that was frankly outlandish, certainly in bad taste for a man of Montaigne's rank, and strictly speaking, quite impious. He had a natural caution about discussing his religious views. He could be extravagant in certain kinds of self-revelations. He wrote about the size of his sex organ, the frequency and complexion of his bowel movements, and the state of his kidney stones with relish. But in matters most intimate—his relationship with God—he kept his counsel.

THE MIRROR OF THE PAST

Montaigne's essays found a delighted audience of readers, though it is more accurate to say that he created his audience by giving them something new, something they did not even know they wanted. Through his writings, contemporaries could discover a different past, filtered through his consciousness, a past that—even if full of unredeemable pagans—was a place of greater safety for exploring the human condition than contemporary Christendom. The mirror Montaigne held up for self-examination was fashioned from the classical past, but what readers could see in it was a clear picture of the present. The massive resurrection of Greek and Roman literature—literature that was utterly orthogonal to the sacred texts of Christianity—gave Europeans an alternative past to claim as their own. Montaigne's innovation was to present this universe to his readers not through pedantic commentaries turgid with learning, but in this experimental form that followed the mental meanderings of a man like themselves—full of faults, complaints, humor, and pity.

What matters for us here is not what the pagans said in particular, but that Montaigne turned more often to the ancients than to the church fathers. He does not ask what Jesus would do. His reference points are more likely to be Plato, Pindar, and Plutarch. He knew his readers would be familiar with the whole tribe of vivid and quotable characters from the ancient Mediterranean who offered many different models from which to learn the art of self-fashioning—or if not familiar, would want to appear so. They could read Montaigne as we read book reviews to gain familiarity with titles we may or may not read. The past became a real time, populated by real people living in places that can be found on maps. Reflected in Montaigne's mirror from the past, readers were given an alternative vision of how life *could* be lived because it revealed to them how life *was* lived. By tapping into the collective memory of humanity, Europeans were in essence resurrecting and revivifying the heritage of an extinct culture, adapting it for their own purposes. Like farmers looking to strengthen their weakening vineyards and orchards, the Europeans grafted an ancient strain of cultural memory onto their own cultural rootstock, diversifying the monoculture of Christendom to fortify it against internal forces of corruption, decay, and sectarian dissension.

Montaigne died in 1592 from complications brought on by his kidney stones. In the following centuries, the classical past loomed larger and more influential with every decade. Through the diffuse vectors of cultural osmosis, it became a prestigious model for artists, writers, and composers. It also provided political paradigms, good and bad, of democracy, republicanism, empire, and of course how quickly all of the above degenerate into tyranny and armed conflict. The publishing industry—and it did begin to look like an industry as printers pushed to secure income streams through new laws granting copyright privileges—accelerated the pace and reach of ideas.

Information inflations produce distinctive shock waves. For the first time, because the excavation of classical literature and the proliferation of printing presses were so fortuitously self-documenting, we can clearly pick out common features of the newly disturbed landscape of memory. New media, be it print, audio recording, or digital cartography, always spawn new writers working in new genres for new audiences, struggling to get attention, throwing in sensational effects to grab eyeballs. Translations of ancient philosophers, historians, and poets such as Aristotle, Caesar, Suetonius, Lucretius, Virgil, and Ovid appealed to a small though influential group of readers. More to the taste of most readers were sensational accounts of travels to strange lands across the Atlantic Ocean, replete with illustrations of unimaginably strange flora and fauna; or the picaresque tales of scoundrels, ogres, and fair maidens told by Rabelais and his imitators. Above all, people read the Bible, recently released from the prison of ecclesiastical Latin and made available in a variety of common spoken languages. The Bible, as those who read the Old Testament know, is full of sex, violence, floods, famine, plagues, family melodramas, miracles, ghosts, apparitions, burning bushes, and assorted scary bits—a full menu of popular entertainment.

The second characteristic of a disturbed landscape of memory is extreme disorientation. All the signposts and road markers have been tossed about or simply buried under the waves of unfiltered information. Once the accustomed barriers to reading and writing are gone, words and images circulate freely, chaotically, and people inevitably experience a kind of vertigo, unable to sense up from down, front from back, past from future. The borders between what is true and what is false become fuzzy. This effect did not even require that printed books reached readers. The ideas almost floated free of the pages and wafted about in the wind, spreading casually in conversations at the market and the local well like stories swapped over water coolers among bored office workers. In Montaigne's day, true facts and made-up facts traveled well beyond the boundaries of the small and coherent urban centers that produced books. Through print, ideas and opinions acquired a physical life of their own and a passport to travel. Made-up facts were particularly powerful agents of opinion making and persuasion. The mere fact of print carried with it an implicit imprimatur of authority. (You can picture them asking each other: "If someone went to the trouble and expense of printing it, it has to be true, right?") Quite contradictory notions of what constitutes value, human dignity, sin, and salvation were released into the oceans of the world's information and created waves of confrontation, competition, and confusion among the reading public and everyone they came into contact with. Print generated buzz: No one needed to read to be influenced by ideas in books.

Montaigne was keenly aware that there was as much humbuggery and pedantry to be found in the books he read as wisdom.

> I have known books to be made out of things never either studied or understood, the author entrusting to various of his learned friends the search for this and that material to build it, contenting himself for his part with having planned the project and piled up by his industry this stack of unfamiliar provisions; at least the ink and the paper are his. That, in all conscience, is buying or borrowing a book, not making one.

But his chief complaint was not against the merciless cobbling together of redundant and inane books in the cut-and-paste fashion we are familiar with today from blogs and sites that compile lists and hyperlinks. He most disliked books that harangued the reader and gave voice to censure and intolerance. He refused to follow this example, making great show of his skepticism and doubt. "My ignorance will excuse me," he half-boasts to his readers. "In general I ask for books that make use of learning, not those that build it up."

In such times, when the confusing circulation of contradictory ideas and so-called facts cry out with equal force to be given credence, we face a crisis of authority. As authorities and institutions fail, we are forced to decide for ourselves which sources are trustworthy and which are not. The question of *what* to believe becomes, almost imperceptibly, a question of *who* to believe. Because once there is this much information swirling around us,

we turn instinctively to the individuals and groups whose authority to speak on the matter seems most trustworthy. We turn to
friends. And when they fail us, we turn to experts and hope for the
best.

Montaigne did not fight his times, but he did refuse to recognize any authority other than his own experience and reason.
"*Que sais-je?*" What do I know? After years of self-interrogation
faithfully recorded in his essays, he concluded that in fact he had
little certainty about the world or about himself—other than the
constancy of change. He came to see life as a state of becoming,
not being. Everything is in flux. He, too, is never fixed. "I aim [in
my essays] only at revealing myself, who will perhaps be different
tomorrow, if I learn something new which changes me." That is
how he came to perfect the art of living amid uncertainty. Memory
provided for him a sense of continuity across time, allowing him
to bend and change rather than break under the extraordinary
pressures of his turbulent times. In the end, rather than learning
how to die well, he learned to live well.

We see in Montaigne's essays early evidence of the power of
direct access to a multitude of texts. For the first time, the collective
memory of the West had expanded beyond the monoculture of
ecclesiastical thought and sacred texts. The diversity of human
experience was revealed to Europeans through the recovery of
their own classical past. And then, there was the ongoing discovery
of new worlds across the Atlantic Ocean. In the century following
Montaigne's death, literacy rates in Europe and the British Isles
grew apace, though at different tempos in different places, concen-
trated chiefly among males in urban areas. But the ideas carried
by books, pamphlets, and maps reverberated well beyond the
circle of those who could read to include those who were read to
in their homes, in the pews of their local church, and in the
growing number of public houses ("pubs") that served alcohol
and the wildly popular imports from the New World and the
Indies, coffee, cocoa, and tea. A new genre was invented—the
newspaper—to meet the growing appetite for novelty, information,
and gossip.

With the advantage of hindsight and knowing that the Enlightenment is to follow, we can detect the idea that knowledge we acquire from the books of our choosing is a path to autonomy, a way to be free from blindly heeding authority. The power of the printed word took on its own aura of truth and righteousness, and here was born the notion that the Bible was literally true. Such an idea was not possible before scripture had been translated from Latin into the common tongue. And in classic pushmi-pullyu fashion, people began to see many different things in the same lines of scripture and headed off in completely different directions. In similar fashion, today the sheer speed of communication across the Internet has accelerated a race to the extremes of intolerance between fundamentalists who urge a return to roots and futurists bent on accelerating the pace of change.

While Montaigne was a public servant and political man until his final years, his sole ambition was to know himself. He reveled in the public acclaim that came from his bestselling books. But he never anticipated, let alone intended, that anything he wrote would shape the course of history or influence the mind of politicians, as Machiavelli intended when he published *The Prince*. It would take another revolution in thinking, the Enlightenment, to canonize the idea that knowledge has the power to change human destiny. Montaigne's library served his private purposes. In the Enlightenment, libraries took on public and very political purposes as well.

In Montaigne's last essay, fittingly called "Of experience," he tells us that "there is no desire more natural than the desire for knowledge. We try all the ways that can lead us to it. When reason fails, we use experience." In the seventeenth century, the quest for knowledge becomes the epicenter of a culture that raises logical and scientific reasoning to new heights, builds instruments to investigate the natural world, and bases its reasoning about Nature on empirical facts. In the eighteenth century, the educated elite came to see reason and experience as the primary tools for building the high road to progress, freedom, and the well-being of mankind. The world of unified Christendom that had failed so spectacularly in Montaigne's lifetime was reborn into a continent of nation-states with vast colonial empires where these ideas were transplanted to native soils. By the time Thomas Jefferson was born in 1743 in a far distant tobacco-growing colony of the British Empire, curiosity had shaken off the noxious reputation it gained in the Garden of Eden. Jefferson grew up believing that curiosity was a natural desire, and to enlightened minds, whatever was natural was good.

CHAPTER FIVE

THE DREAM OF THE UNIVERSAL LIBRARY

If a nation expects to be ignorant and free in a state of civilization, it expects what never was and never will be.

—THOMAS JEFFERSON TO CHARLES YANCEY,
JANUARY 6, 1816

THOMAS JEFFERSON (1743–1825) knew Montaigne's essays well and shared Montaigne's passion for Cicero. But he was a political man, and when he read Cicero, it was for political advice, not lessons about how to live well in the face of certain death. Jefferson was studying how to move his countrymen from being subjects of a king to citizens of a republic. Having won the war of liberation from the British monarch, he turned quickly to the business of forming a new man, a free enlightened man, an American citizen. There had not been a free citizen of a republic since Cicero's time. History would be the sole instructor.

For men and women of the Enlightenment, knowledge was something to be acquired, organized, and shaped into an instrument of progress. The Founders believed ideas freely shared among people of goodwill would speed the process of enlightenment. In a letter to Isaac McPherson on the matter of patents and property, Jefferson wrote that "an idea, the fugitive fermentation of an individual brain" cannot be the exclusive property of one.

> Its peculiar character, too, is that no one possesses the less, because every other possesses the whole of it. He who receives an idea from me, receives instruction himself without lessening mine; as he who lights his taper at mine, receives light without darkening me. That ideas should freely spread from one to another over the globe, for the moral and mutual instruction of man, and improvement of his condition, seems to have been peculiarly and benevolently designed by nature, when she made them, like fire, extensible over all space, without lessening their density in any point.

In this prophetic vision, Jefferson describes a virtual network that transmits ideas across time and space, without any natural barriers—except those placed by men claiming ownership. (His letter is an extended argument against long copyright or patent terms and other impediments to the free flow of ideas.) We recognize this network as the Internet. The centuries of technical innovation from Jefferson's day to the present have been energized by a natural desire for knowledge that could be put in the service of the moral instruction of man and the improvement of his condition. These aspirations were born in the Enlightenment and are fully visible in the life and thought of Thomas Jefferson. This new faith in progress valued knowledge for its instrumental purposes—its ability to change the world, influence people, and secure happiness and liberty for all. Ever an impatient dreamer, Jefferson tried to hasten the arrival of the future and championed the power of organized knowledge to change the world. He dreamt of a library that would provide unfettered access to the sum of human knowledge, to be grown from the seed of his own library in Monticello. Ironically, war and fire played the part of creator, not destroyer, of a library for perhaps the first and only time in history.

KNOWLEDGE AS AN INSTRUMENT OF POLITICS

In late April 1813, Yankee soldiers invaded the capital of Upper Canada, York (present-day Toronto). They burned down the parliament building, and with it imperial records and the parliamentary library. In war as in peace, one good turn deserves another. In August the following year, British soldiers returned the favor: They set fire to the U.S. Capitol and destroyed the congressional library.

To replace its stocks, Congress bought the private library of Jefferson, then living in retirement at Monticello. They acquired his collection of 6,487 volumes comprising 4,931 unique titles—the books, maps, manuscripts, and musical scores that in their day made up the largest, most diverse, and most comprehensive collection of recorded knowledge in the Western Hemisphere.

So, the story goes, it was with this casual act of vengeance that the dream of a universal library was born in America. The dream was a simple one: all human knowledge brought together in a single collection and made accessible to all for the general enlightenment of the people. This creation story has come to define the culture of knowledge in America, where it is held that the enlightenment of the people will invariably lead to democracy and wise self-rule. As Thomas Jefferson wrote, "Enlighten the people generally, and tyranny and oppressions of body and mind will vanish like evil spirits at the dawn of day." Jefferson, America's most devoted bibliophile and hyperopic optimist, became the patron saint of American libraries.

In the seventeenth century, books arrived on American soil already sanctified by divine purpose, so it is not surprising that they assumed a lofty status among the Founders. The first libraries were collections of religious books, packed in the luggage of religious pilgrims from the Old World. The Protestants who settled the northern colonies were people of the Good Book. Scripture and theological works were the primary furniture of their spiritual lives. These collections had one purpose—to impart knowledge of God and his ways. And they were read with one goal in mind—personal salvation.

By the eighteenth century, the Republic of Letters—an imaginary community of enlightened people whose spiritual homeland was wherever they dwelled—had expanded from continental Europe to the distant shores of the Atlantic and began to encroach on and eventually crowd out the zealous. Continental ideas about enlightenment and secular learning came into port with books from Europe, and there were increasing numbers of colonists who assembled libraries for quite secular purposes, including agriculture and trade, governance, political enlightenment, and entertainment. Jefferson's was by far the largest, but he lived among men and women who shared his love of books and learning.

For Jefferson, the goal of reading was not salvation but freedom. By his lights, to live as a free man one had to possess knowledge. After the colonies achieved political independence from Great Britain, Jefferson and his peers grappled with the great responsibilities that freedom entails. They believed these responsibilities could only be borne by an informed citizenry, with knowledge of current events, trade and agriculture, history and political theory, and above all, knowledge of themselves.

These were moral as well as civic purposes, and the Republic decided early in its life that the moral sphere could thrive in civic life only in a nondenominational way, with no state religion and no privileged clerisy. We hold that democracy demands equal access to goods, services, and knowledge. The culture of knowledge in America has been a servant of democratic governance. This instrumental view of knowledge meant that three principles would become fundamental to American-style democracy: The press must be free, the government must be open and accountable to the people, and the education of the citizenry must be a right and responsibility of the governed and their representatives.

Libraries became vital instruments in guaranteeing all three. But it was not always so. And it was far from clear that in order for libraries to perform these obligations they would need to provide access to a universal collection. That was pure Jeffersonism. When Congress set about reconstituting the collection of law books, treaties, and maps that perished in the War of 1812, they were not seeking to assemble a universal collection. But Jefferson had his own reasons to offer his library for sale as what he modestly called "a replacement."

JEFFERSON THE GRAND ACQUISITOR

Thomas Jefferson knew personally what it was like to lose a library to fire. In 1770, when he was twenty-six years old, his ancestral home in Shadwell, Virginia, burned and with it, his library of several hundred volumes. He quickly began to assemble another collection, and within three years he recorded in his notebook that he had a total of 1,256 volumes at his new home, Monticello. He no doubt had more than this, as these notes record neither his music collection nor the books he had in his office in Williamsburg.

Jefferson was more than a man of learning who loved his books. He was a self-confessed bibliomaniac—a grand acquisitor—and he used every possible opportunity to pursue the purchase of more books. He thought about books, he dreamt about books, he drew up lists of books that he desired, and wherever he found himself, no matter how busy he was, he contacted book dealers and set up standing orders for the purchase of what was on his desiderata list. He enjoyed the chase so much that during his years in Paris, where from 1784 to 1789 he represented the newly independent states as commissioner and minister to France, he made fast friends with book dealers throughout the continent and the United Kingdom. He was one of their best customers, collecting not only on his own behalf, but also for friends such as James Madison and James Monroe. "While residing in Paris, I devoted every afternoon I was disengaged, for a summer or two, in examining all the principal bookstores, turning over every book with my own hand, and putting by everything which related to America, and indeed whatever was rare and valuable in every science." He had not brought his entire collection with him to Paris, and this was an excellent reason—if he needed one—to buy second copies of books he liked. In 1789, for example, he picked up another edition of Montaigne's essays, though he already owned the title.

His library also grew by buying up other people's libraries, including those of Benjamin Franklin and fellow Virginians Richard Bland and Peyton Randolph, who shared his interest in the early history of their state. Purchasing libraries is a common practice among bibliophiles. Comprehensive collections are frequently assembled by harvesting the fruits of other collectors' zeal. But buying books wholesale from existing collectors was particularly notable among monarchs and statesmen of start-up countries. The emperors and empresses of Russia, for example, eager to catch up with centuries of European intellectual achievement, were famous buyers of libraries, from Peter the Great's purchase of Archibald Pitcairne's library on the art of physic (medicine) to Catherine the Great's acquisition of philosophe Denis Diderot's library.

With a double-mindedness characteristic of Jefferson as a man—and typical of collectors and other addicts—he found in the larger ambitions of his collecting a justification for spending more money than he had in the pursuit of his quarry. That larger ambition was to create a bank of knowledge full of the accumulated riches of human thought and memory that the young nation could draw on for generations to come. His acquisitions, then, were driven by two overlapping but sometimes conflicting ambitions. He wanted a collection for his private purposes, shaped by his own personal bibliographic tastes, to be housed at Monticello and enjoyed in domestic intimacy. And he was seeking to build a library that would enshrine his political and intellectual ideals and in good time be put in service to the nation. As he was leaving public office behind for good in 1809, he wrote to Madison that he contemplated giving his collection to a university—either to Virginia, or to the national university that might be built in the capital.

But war intervened, and so, too, did a personal financial crisis. While officials in Washington had ample warning of the British invasion of 1814, there were only desultory evacuations of the books and papers housed in the Capitol, with priority given to government records over books from the congressional library. The clerk of the House (who also held the position of librarian of Congress) was absent from his post, away at a spa in Virginia taking the waters to fortify his constitution. The staff he left on duty did the best they could to find conveyances to move the books out of harm's way. But most horses and carriages had already been commandeered by the combatants. As a result of the Capitol fire, countless books, maps, records, and vital documents (including Revolutionary War pension claims) were lost. It is not clear how much of the three-thousand-volume congressional library remained when the fighting was over. The reports of widespread depredation suffered by the Capitol were later judged to be exaggerated. What flames there were, were used by propagandists to fuel the anti-imperial patriotic fervor of the young Republic's citizens. Contemporaries gave eager ear to the rumors that British troops deliberately pulled books, manuscripts, and maps from the library in order to stoke the fires.

Meanwhile, Thomas Jefferson had his own reasons to believe the worst. Living in retirement at Monticello, he was a man with many books, many expensive habits, and many debts. Chief among those expensive habits was Monticello, an ambitious building project that was ongoing for most of his life. "Architecture is my delight, and putting up and taking down, one of my favorite amusements," he confessed. And he kept himself very well amused. Within a few weeks of the fire, Thomas Jefferson approached Congress about purchasing his library to replace what was lost. Through an intermediary, he presented a formal tender to the Library Committee of Congress. But his offer was driven as much by his ambitions for the young Republic as by pecuniary interests.

Jefferson had always cared deeply about the congressional library: As president (1801–1809) he formalized the office of librarian of Congress, ensured a dedicated acquisitions budget, and took active interest in the growth of the library. In his tender, he enclosed a catalog of his books and invited Congress to name their price. Congress offered him $23,950, a price based solely on the size and number of the books—$1 for a duodecimal, $3 for an octavo, $10 for a folio, and so on. The Senate, with typical liberality (the House appropriates funds, not the Senate), quickly passed a bill approving the purchase. But not the House. They had a far more partisan view of the former president, and they had the power of the purse. A contemporary report states:

> Those who opposed the bill, did so on account of the scarcity of money, and the necessity of appropriating it to purposes more indispensable than the purchase of a library; the probable insecurity of such a library placed here; the high price to be given for this collection; its miscellaneous and almost exclusively literary (instead of legal and historical) character, &c.
>
> To those arguments, enforced with zeal and vehemence, the friends of the bill replied with fact, wit, and argument, to show that the purchase, to be made on terms of long credit, could not affect the present resources of the United States; that the price was moderate, the library more valuable from the scarcity of many of its books, and altogether a most admirable substratum for a National Library.

There were some, such as Cyrus King of Massachusetts, who tried to exclude "all books of atheistical, irreligious, and immoral tendency," of which there were quite a few, given Jefferson's broad interests and special fondness for philosophes. But careful selection demands the luxury of time, which the members of Congress did not have. In the end, the measure was approved along strict party lines.

So the first chapter in this bibliographical genesis ended happily for all: Congress had acquired a library, and Jefferson paid off his major creditors. In the end, he even pocketed $8,580—a sum it turns out could buy a lot of books and wine.

JEFFERSON THE STATESMAN

What exactly had Congress bought? Some members were clearly already thinking about "a National Library," and they saw this

purchase as a good foundation. Jefferson himself was less ingenu-
ous. In pitching the sale to Congress, he insisted that the books
were each and every one of them necessary for Congress.

> I do not know that it contains any branch of science which
> Congress would wish to exclude from their collection; there is,
> in fact, no subject to which a member of Congress may not have
> occasion to refer . . . [The collection], while it includes what is
> chiefly valuable in science and literature generally, extends more
> particularly to whatever belongs to the American statesman.

And there were indeed a lot of useful books for the making of
good laws—the great English jurists Blackstone, Bolingbroke, and
Coke in particular—in addition to dozens of volumes on foreign
and domestic laws, treaties, books on weights and measures, trade
and customs, and so forth. And because the congressional library
was also the only library accessible to members and government
officials in the cultural backwater that Washington was, who
would deny that the majority of books, while not precisely
important for the purposes of legislating, would not be welcome
as companions to the soul? Gentlemen of the capital could settle
down with the wisdom of Montesquieu, Voltaire, Locke, and
Hume. Members could choose to pass a few hours in the company
of the great poets and historians of the past—Homer and Virgil,
Ovid and Xenophon, Thucydides and Herodotus, Tacitus and
Cicero, Dante and Shakespeare, Milton and Ossian (whom
Jefferson deemed "the greatest poet who ever existed," though
we now know he did not actually exist). And for those lighter
moments, between writing legislation and the weighty deliberations over treaties, members could tuck into delights of literature
by the likes of Sterne, Defoe, Cervantes, and Rabelais.

But the idea that each volume in the library was somehow
germane to governing was a flagrant exaggeration, a high-minded
sales pitch that only Jefferson could have made with a straight face.
What was true was that Jefferson believed his library contained
what it took for Americans to educate themselves, to advance

science and philosophy, and to form their identity as Americans distinct from all other nations.

Jefferson was a futurist. That was why he was obsessed with the past. "It is the duty of every good citizen," he wrote, "to use all the opportunities, which occur to him, for preserving documents relating to the history of our country." He was not an antiquarian, reveling in the discreet charms of the simple and obsolete modes of life lived by generations long gone from the face of the Earth. The purpose of saving the historical record was to discover in it the ways in which America differs from Europe—is as good as Europe, in fact probably better, but certainly not inferior. What it is to make an American citizen out of a British subject, Jefferson argued, is the moral, physical, aesthetic, and political environment in which the republican citizen is raised. The clean air, unpolluted waters, rich soils, and vast expanses of the continent allow humanity the greatest range for the development of its higher self, inspired and instructed by the accumulated examples of the past.

He was first and foremost interested in items that documented the nation: everything written by Americans and everything written about America. Jefferson collected in all the publishing formats available in his time, ranging from books, newspapers, and broadsides to maps, charts, manuscripts, engravings, statistical tables, and musical scores. His interests went far beyond politics, public policy, and political economy. There were works by American scientists and poets (including the first black poet, the ex-slave Phillis Wheatley), works about the natural history of the continent and the languages spoken by the indigenous populations. In a letter thanking James Madison for sending him a pamphlet on the Mohican language, he wrote, "I endeavor to collect all the vocabularies I can of the American Indians, as those of Asia, persuaded that if they ever had a common parentage it will appear in their languages."

But in Jefferson's day, the world of learning lay elsewhere, so the majority of his books came from Europe and a good portion of them were in European languages—Greek and Latin, German, French, and Spanish. He collected histories of the classical world,

European history, and everything that would provide information about the republican predecessors of the United States. As he wrote in the preamble to the education bill he drafted for the Virginia legislature, A Bill for the More General Diffusion of Knowledge, Jefferson especially believed a future secure from tyranny and corruption demands that leaders "illuminate, as far as practicable, the minds of the people at large, and more especially to give them knowledge of those facts, which history exhibiteth, that, possessed thereby of the experience of other ages and countries, they may be enabled to know ambition under all its shapes, and prompt to exert their natural powers to defeat its purposes." Montaigne, too, advocated the study of history: It brings us into the company of "the worthiest minds, who lived in the best ages."

The record of the past is direct evidence of man's potential, both good and bad. Jefferson owned numerous books he deemed exemplars of bad ideas and pernicious influence. For example, he admired Hume's essays but roundly condemned his history of England, blaming this book and those of Blackstone for the spread of "universal toryism over the land." And yet, though he blamed these books for spreading bad ideas, he did not purge them from his library. A universal library must include all books of significance, whether their influence be for good or ill. Having access to recorded knowledge and a reliable record of the past—feats and follies, virtues and vices alike—became a lynchpin of self-rule. An enlightened people will and must judge for themselves where truth lies.

So how much can this library tell us about the mind of Jefferson? Access to the originals used by an individual can be revealing, especially if the books bear annotations. They can be like magnifying glasses held up to the mind, because reading the marginalia—the comments, quarrels, notes, and doodles we find in the margins or fervidly squeezed in between the lines of a page—is like hearing the owner whispering in your ear. But Jefferson, as a rule, did not write in his books. He tended mainly to make owner's marks—revealing in itself, but not necessarily about his intellectual concerns. We cannot see or hear him arguing with a writer, as we see Ben

Franklin doing in furious annotations in his books. One book that Jefferson had acquired from Franklin's library had annotations covering the page and into the gutters, memorializing Dr. Franklin's passionate indignation at the stupidity and mendacity of the views therein expressed ("An impudent falsity," "Another misrepresentation," "This is a most extravagant assertion," and so on).

Jefferson was more an acquisitor than an annotator. He dedicated himself to assembling a universe of learning, neither commenting on it nor quarreling with it. To read his mind, we do better to study his correspondence than look inside the volumes on his shelves. In his epistolary writing he carried on lively debates with others—and with himself—as his letters are rich with references to writers, thinkers, and historical actors. Even lightly browsing his bookshelves will yield more knowledge about the man than would the same time spent carefully studying each volume. There were no doubt items in his library that he did not, in fact, read— the three-volume book on statistics in Russian stands out as one example—but the fact is of little consequence. That he thought it important to collect such an item reflects his view of the world of knowledge and casts light on his dreams of the future of the United States. It is this notion—that of a comprehensive and coherent collection of knowledge about and useful to America, not merely catering to personal interests or executed in imitation of aristocratic models of collecting—that makes people describe his collecting practices as universal and stand as a model for the Internet today. But this vision also reflects the essential character of Jefferson's spirit—omnivorous, insatiable, and quixotic.

JEFFERSON'S LAST ACT

The paradox for a man like Thomas Jefferson, both a diligent student of history and one who self-consciously made history, is that no matter how much we have to learn from the past, in making history we inevitably obliterate much of what came before us. "The earth belongs to the living and not to the dead," Jefferson wrote to Madison from Paris in the early months of the French Revolution. Revolutionaries—and Jefferson was nothing if not a committed revolutionary to his dying day—engage in acts of destruction that they hope are creative. They set out to destroy what is old and create what is new. But in any successful social or political revolution, the originators inevitably lose control of both processes. Jefferson was no different. And yet he believed the national library would endure.

Jefferson's rationalism and native idealism waxed and waned as he lived longer than he had expected. By the time he died in 1826—exactly fifty years to the day since the signing of the Declaration of Independence he drafted—he witnessed developments in the United States that were completely counter to what he had hoped and indeed planned for. Where he had envisioned a largely agricultural nation spread evenly across the capacious land, he saw instead the steady influx of migrants, the growth of manufacturing, and the burgeoning of cities. As he fought off the pain of disillusionment, he withdrew from the present and gave himself over to his utopian longings for a future that would be all peace and harmony among men. Like his fellow survivor John Adams, Jefferson was very concerned that history should remember him for the role he played in the founding of what he expected to be a great nation. By the time Jefferson and Adams lived into their eighties, they were afraid that they would not be remembered for the things that they had accomplished and worse—that they would not be remembered at all. It was as if they had a bitter foretaste of America's inane tension between nostalgia and utopian futurism, ignoring altogether the unadorned facts of history and the lessons these men so hoped to leave behind to instruct their descendants.

This is where Jefferson's library could perform its final act of redemption. Just weeks after his books left Monticello for the capital in 1815 and the shelves of his library stood bare, Jefferson wrote to Adams that "I cannot live without books; but fewer will suffice where amusement, and not use, is the only future object." He had discharged his duty as a public servant and desired nothing more than to enjoy the privileges of being a private citizen. And so, bereft of books and debts retired, he entered in the third chapter of his bibliomania, the creation of his last library.

He started assembling it right after the sale of his library to Congress, and this one he designed to be a library for a retired man, not a working library. For while Jefferson confided to his fellow revolutionary and comrade-in-arms that he was more comfortable with his dreams of the future, to another friend he wrote that "I feel a much greater interest in knowing what has passed two or three thousand years ago, than in what is now passing." So his third library was stocked, like Montaigne's, primarily with the classics, particularly Greek and Latin histories. We have the testimony of his family that these volumes were his constant companions in his retirement. At the same time, he was occupied with planning to translate his library to the university that he was creating in Charlottesville. Again succumbing to the passion of collecting and dreaming of the future, he thought about building a comprehensive collection with a coherence that could structure a curriculum for the citizens of the newborn republic. He reportedly drew up a list of the books that should be housed under the great dome of the rotunda he designed for the campus. It numbered 6,680 volumes, he reckoned, and would cost $24,076.

At the time of his death in 1826, Thomas Jefferson had over two thousand books in his library. These books were to go to his university in nearby Charlottesville. But he was improvident to the end, and his heirs were forced to sell his books to pay his debts. The University of Virginia, rather than receiving this bequest, was instead the recipient of an embarrassed letter from his grandson and executor explaining that the books would go, along with other movable property (including slaves), to his creditors.

But Jefferson's vision of libraries is very much alive. In a country that has granted greater prestige and accommodation to private property than to public goods—even in the realm of ideas—American libraries became sanctuaries for a shared ambition of self-improvement, like great public gardens of the mind. They serve an individual, a community, or a society as a central gathering place where you can find all the fruits of civilization, the harvest of a cultivated mental landscape that has been tended for generations. True to the Jeffersonian model, every great library has its fair share of exotics—things that surprise, delight, scandalize, or amuse because they testify to the essential idiosyncratic nature of humans and their interests. A great library is always well stocked with extinct species of human thought and creativity, cultivars that were created generations ago and preserved from societies that have been lost to time. All these fruits of human labor and imagination are worth care and keeping, because, as Jefferson suggested, in a democracy there is no subject of which we might not need knowledge.

Thomas Jefferson gave Americans the ideal of a universal collection that must include the unique, the quirky, the nonconforming—the "free radicals" in society that Americans like to think of as agents of innovation and progress. He and his peers also gave us the idea that free access to information is the sole guarantor of self-rule.

HEROES OF TIME

The Founders mandated that access to information be guaranteed in a democracy. Subsequent generations backed that mandate up by building publicly funded institutions to provide that access to multiple generations. Where the government did not assume full financial responsibility for schools and libraries, its successors in Congress created financial incentives for private groups to support them. Further they mandated that in a democracy these institutions should be dedicated to the public good and have political autonomy, protected by the government and supported by the public. As Jefferson knew, private collections are the vanguard of our collective memory, but their value is realized only when they pass into the public sphere.

Although libraries are perennially cash-strapped and struggle to create a public image that will attract donor dollars and public support, the mission of libraries is best served by maintaining a low profile, even at the risk of appearing unglamorous and stodgy. They need to be protected from the whims of private taste and demands of political expediency. As Daniel Kahneman has pointed out in another context, institutions are important for functions that must persist over long durations of time. Their job is to slow us down, to add friction to the flow of thought, foster inertia, and carve out from the fleeting moment a place for deliberation. But because they move slowly and keep their eyes fixed on the long horizon, cultural institutions can miss the action happening on the ground, right before their eyes.

That is why collectors are so valuable a part of an information ecosystem. The faster the present moves, the more valuable they become. Collectors historically have acted as the first-line defense against the physical loss of our cultural legacy. They collect and curate the artifacts of knowledge on our behalf. While the motives of individual collectors can vary between the poles of intellectual curiosity and personal vanity, great collectors have some larger purpose they wish to accomplish and to which they dedicate enormous amounts of time and treasure. They are the ones who keep the strategic reserve of memory rich, saving various fossils of extinct cultures and ensuring that the collective memory of mankind does not become a monoculture. Collectors are heroes of time, marrying their private passion with a public purpose, foreseeing what will have value tomorrow but can only be collected and preserved today.

Over the nineteenth century, as libraries came to assume public purposes and receive public support, the relationship between the individual collector and political society changed irrevocably. Public libraries serving local populations grew in scope and value, and individuals such as Andrew Carnegie who pursued private wealth returned some of that treasure through building libraries for the public. At the same time, libraries never lost their sense of being private sanctuaries, like Montaigne's tower. Even in the age of large, publicly accessible libraries and the even larger and more accessible Internet, the library lives in our imagination as a private place. Each of us assembles our own library over the course of our lives, the collection of all the bits of information, knowledge, music, and art that we cannot part with. Whatever their form and wherever we keep them—paper or computer chip, on a bookshelf or in the cloud—they are the things that we call our own, to use and control as we wish. With dismay and fear, we are learning how easy it is to invade private spaces online and how little control we have over our own content once it is online.

Thomas Jefferson has deeded to us a vision of the library comprehensive in ambition, if not literally in scope, and organized for use. When web natives go online, they expect to find a universal collection of knowledge. As far as they know, if it is not on the web, it does not exist. Google founders Sergey Brin and Larry Page (the latter the son of a librarian) aspire to "organize the world's information." The technology entrepreneur Brewster Kahle has created the Internet Archive to ensure "universal access to all knowledge." Dozens of libraries, public and private, have come together to build the Digital Public Library of America, designed to become a virtual front door to the collective holdings of libraries across the country. Europeana is yet more ambitious. Over two thousand institutions have contributed digital scans of books and manuscripts, archives, films, paintings, and sculptures in an effort to broaden access to European culture. Before the Enlightenment, the idea of a right to access, let alone to a universal collection, would have been nonsensical. Now it is the default expectation.

ON CHRISTMAS EVE OF 1851, a quarter of a century after Jefferson's death, fire again destroyed the Library of Congress. More than 35,000 volumes perished, including two thirds of the 6,487 volumes in the original Jefferson collection. The library room in the Capitol building was rebuilt the following year in cast iron. And within two decades, after the Copyright Office was moved into the congressional library, the shelves began to groan under the weight of copyright deposit books, journals, sheet music, and other records of the creativity of the American people. The library's holdings reached scales that were previously unimaginable. The collections eventually required the building of three immense edifices on Capitol Hill, with collections also housed in secondary storage buildings in Maryland and Pennsylvania. And in 2007, the library opened a brand-new center for its audiovisual collections, among the richest yet most ephemeral records of life. A high-security Federal Reserve storage facility, decommissioned after the Cold War, was converted into a center for the library's collections of over 1.1 million film, television, and video holdings, together with 3.5 million sound recordings. The Library of Congress Packard Campus for Audio-Visual Conservation is in Culpeper, a small town in Northern Virginia, just off the road to Monticello.

PART TWO

WHERE WE ARE

In Los Alamos, I met physicists and other "natural" scientists, and consorted mainly, if not exclusively, with theoreticians. It is still an unending source of surprise for me to see how a few scribbles on a blackboard or on a sheet of paper could change the course of human affairs.
—STANISLAW ULAM, PREFACE TO HIS MEMOIR
ADVENTURES OF A MATHEMATICIAN, 1976

CHAPTER SIX

MATERIALISM: THE WORLD IS VERY OLD AND KNOWS EVERYTHING

—Say it, no ideas but in things—
—WILLIAM CARLOS WILLIAMS, "THE DELINEAMENTS OF
THE GIANTS," *PATERSON*

WHEN JEFFERSON RETIRED IN 1809, he wrote to a friend that "Nature intended me for the tranquil pursuits of science, by rendering them my supreme delight. But the enormities of the times in which I have lived, have forced me to take a part in resisting them, and to commit myself on the boisterous ocean of political passions." Jefferson fully expected that after the exertions of the revolutionary generation to establish the Republic, the tides of political passion would ebb. The advancement of science and learning would ensure political stability and economic prosperity. But Jefferson's belief that the pursuit of knowledge would keep enemies from the pursuit of each other was proven false in the testing of it.

He wrote this letter two weeks after the birth of Charles Darwin and Abraham Lincoln. (Both were born on February 12, 1809.) In the space of just fifty years, the "enormities of the times" had not diminished. On the contrary, the struggles to secure liberty and the pursuit of happiness for all led the country to the brink of civil war. In 1859, Lincoln entered the presidential race, setting sail on the high seas of political passions and marking a course to a port from which there was no return.

And in 1859, science, rather than continuing tranquilly in the pursuit of truth, on the contrary entered a fateful and fractious boom time that continues to transform the daily lives of people across the globe. Darwin, who had spent decades quietly pursuing his studies of barnacles and pigeon breeding, rushed *On the Origin of Species* into print. Fully aware of the calamitous implications of evolution for the understanding of human origins and struggling with intense anxiety about its reception, he had steadfastly procrastinated writing up his theory. Then one day he learned he was about to be scooped. An obscure scientist fourteen years his junior, Alfred Russel Wallace, had reached the same conclusion and was seeking to publish his findings. In the event, they copublished the seminal paper that announced evolution to the world, and Darwin went on to write his own treatment in *The Origin*. Immediately upon its publication, the scientific pursuit of knowledge itself became a boisterous ocean of passions.

Within a few decades of Jefferson's death in 1826, a new science, anchored firmly in the bedrock of materialism, had superseded natural philosophy and natural history as the normative method of understanding the world. As one historian notes,

> The word "science" (from the Latin *scientia*, meaning knowledge or wisdom) tended to designate any body of properly constituted knowledge (that is, knowledge of necessary universal truths), while inquiries into what sorts of things existed in nature and into the causal structure of the natural world were referred to, respectively, as "natural history" and "natural philosophy."

These two arenas were separate and distinct, practiced by two separate groups of individuals, with the philosophers claiming greater intellectual prestige than historians. After the two fields converged in the nineteenth century, it became impossible to understand any material phenomenon or effect without mapping its underlying material cause.

The technologically advanced, data-rich world we live in today all devolves from one central discovery made in the nineteenth century: The universe was created in deep time, extends infinitely across deep space, and *leaves a record of its own history in matter.* The material universe is itself an archive and the Earth writes its auto-biography in matter.

The discoveries produced by the new evidence-based science engendered a permanent revolution of thought about Nature and human nature that is still spinning out today. The definition of what constitutes a material effect has broadened to include all states of Nature, *including mental states.* As a consequence of the scientific revolution that began in the seventeenth century, when men (and a handful of women) of science explained away the mysteries of natural events such as lightning and magnetic fields, people began to see Nature acting according to laws that were accessible to human reason. They increasingly accepted and eventually preferred these explanations to the heavenly interventions and supernatural powers they had routinely deferred to. These ideas undergird all our technologies today, from airplanes to X-ray machines. Whatever our philosophical or religious orientations, these ideas have been assimilated into culture for political, military, and economic reasons as much as for their intellectual persuasiveness.

Rather than viewing the discovery of evolution as an act of a singular genius, we see now it was only a matter of time, once naturalists had come to believe that the world is very old and matter is a form of memory. Gradually but resolutely, they aban-doned the notion that our body of natural knowledge comprised a set of truths revealed through scripture and prophets. They came instead to trust their powers of observation and deduction. In their eyes, the world itself was transformed into a vast landscape of clues to knowledge hidden in plain sight. This perceptual shift gained ground among men of learning obscurely, quietly, unnoticed by the world at large, as certain inquisitive naturalists took to the fields and looked very closely at the earth beneath their feet. As the industrial age got under way in the late eighteenth century, mining engineers blasted through quiet landscapes to carve canals for moving raw materials to manufacturing sites and manufactured goods to commercial centers. As they did so, they laid bare distinctly stratified layers of rocks buried under feet of soil. On close inspection of freshly exposed terrains, the naturalists came to the awkward and shocking conclusion that the rocks and fossilized life forms embedded in them predated the creation of the Earth. Something was wrong here. They concluded that it was the received wisdom that was in error, not the rocks.

We now take for granted that our planet is 4.5 billion years old and the universe 13.75 billion years, give or take. We think that the future will unfold over more billions of years toward an uncertain fate—either a slow and cold paralysis, or a rapid and violent implosion back into the seed of another universe. However it turns out, we will be long gone.

But two hundred years ago, such numbers and notions were preposterous. Then the common view was that the world had a birth date—the days of creation as told in the book of Genesis. And it had an end date—the Second Coming of Christ (TBD). In 1650 Bishop Ussher, primate of all Ireland (1581–1656), published a learned chronology that established with exquisite mathematical precision that the world was created in 4004 B.C. His method of calculation, based on rigorous and painstakingly detailed research into the Old Testament, was widely accepted. This chronology fit nicely into the common view of history as a sacred narrative, progressing purposefully from an intentional beginning toward a promised consummation—in short, a teleology. Like many of his peers, Jefferson subscribed to Ussher's time frame. He was the proud owner of mastodon bones, but he did not recognize them as fossils and he rejected the idea that species ever go extinct.

In Jefferson's day, "science" denoted reason-based inquiry into and systematic knowledge of all subjects, from the origins of rocks and the nature of lightning to the political economy and Homeric poetics. Science, being "reason in action," was hailed as a force for liberation and expansion of the human spirit. As Jefferson wrote, "Freedom [is] the first-born daughter of science." The American Philosophical Society, founded by Benjamin Franklin in 1743 to "promote useful knowledge," "add to the common stock of knowledge," and pursue "all philosophical Experiments that let Light into the Nature of Things, tend to increase the Power of Man over Matter, and multiply the Conveniencies or Pleasures of Life," counted Washington, Adams, Jefferson, and many other political figures among its membership. Jefferson was elected its president in 1797, the same year he became vice president of the United States. As his political adversaries were fond of pointing out, Jefferson was usually as attentive to the duties of the former as he was to the latter. He held this office until 1814, throughout his vice presidency and presidency of the United States, only stepping down when he was in his seventies.

As Jefferson wrote to the president of Harvard, men of their generation "have spent the prime of our lives in procuring [young people] the precious blessing of liberty. Let them spend theirs in shewing [sic] that it is the great parent of *science* and virtue; and that a nation will be great in both, always in proportion as it is free." Jefferson made the pursuit of knowledge a matter of state policy and national defense. "Science is important to the preservation of our republican government, essential to its protection against foreign power." He identified the growth of knowledge with the growth of freedom, but also with the growth of the United States' economy, population, and political territory. Jefferson used his political office to extend not only the borders of the Republic, but also the boundaries of knowledge about the continent and beyond. He commissioned ambitious geographical and scientific explorations and sponsored the collection of artifacts, specimens, maps, and documentation from the far reaches of the continent. As soon as he had secured the Louisiana Purchase, he sent Lewis and Clark off to find out just what was out there (as well as to map it for use in boundary disputes bound to arise between nations). During his residence at the White House, the East Room was fully kitted out with his collection of paleontological wonders. He would show these off to visitors as he discussed the particular features of the flora and fauna of the New World, attending with particular care to the size of things relative to those found in the Old World.

Visitors wrote with awe about exotic live specimens of wondrous American fauna kept on hand to dazzle them. Grizzly bears and a prairie dog, living souvenirs of the Lewis and Clark expedition, graced the lawns of the newly built White House and delighted his startled visitors.

THE FACULTIES OF THE MIND

Even as Jefferson was president, the Age of Reason was yielding quickly to the Age of Matter, when empirical science rigorously based on evidence superseded natural philosophy and natural history. This shift is manifest in the evolution of Jefferson's theory of knowledge, vividly documented in his correspondence and book catalogs, like fMRIs of his intellectual migration to materialism. While he was an exceptional man, his embrace of materialism was very much of its time. Jefferson organized his library according to a practice that dated back to Francis Bacon (1561–1626), one of his idols. The enlightened philosophes in France used the same scheme in their systemization of all knowledge, published in the multivolume encyclopedia.

Bacon's categorization of learning, expounded in *On the advancement and proficiencies of learning* (1605), describes "the Emanations of Sciences, from the Intellectual Faculties of Memory, Imagination, Reason." Bacon maps the internal landscape of knowledge according to these three mental primitives, the irreducible features of the mind from which our intelligence emanates. (We have replaced this understanding with the notion that our minds emanate from physiological features of the brain, of which memory, reason, and imagination are emergent phenomena.) When Jefferson compiled a catalog of the books he was shipping to Congress in 1815, he began by explaining his cataloging scheme.

Books may be classed according to the faculties of the mind employed on them: these are—I. Memory; II. Reason; III. Imagination.

Which are applied respectively to—I. History; II. Philosophy; III. Fine Arts.

His books were shelved in this order at Monticello, and the order was maintained at the congressional library for decades.

But long after he shipped his books off and had begun his third and smaller library, Jefferson continued to research and refine his categorizations of knowledge. In 1824 he writes to a friend that:

> Lord Bacon founded his first great division on the faculties of the mind which have cognizance of the sciences. It does not seem to have been observed by anyone but the origination of this division was not with him. It had been proposed by Charron, more than 20 years before in his book *de la Sagesse* . . . and an important ascription of the sciences to these respective faculties was there attempted. This excellent moral work was published in 1600. Lord Bacon is said not to have been entered on his great works until his retirement from public office in 1621.

Note how precise Jefferson is in dating which author published first: his letter bears the enthusiastic pedantry of the graduate student. Pierre Charron (1541–1603), a contemporary and admirer of Montaigne, drew these categories from what he conceived to be the anatomy of the brain and especially its ventricles, where he believed the soul dwells. Thus a moist temperament produces memory, a dry temperament intelligence, and a hot temperament the imagination.

As Jefferson aged, he came to view humankind as naturalized beings, no different in essence from other higher animals. In an early manuscript catalog of his library, compiled in 1783 when he was forty, he has written a note under moral philosophy (part of ethics): "In classing a small library one may throw under this head books which attempt what may be called the Natural history of the mind or an Analysis of its operations. The term and division of

Metaphysics is rejected as meaning nothing or something beyond our reach, or which should be called by another name."

Over time, the category of Reason or Law in his library thinned out as he reassigned more titles to Memory or History to accord with his changing views. Shortly before his eighty-first birthday, Jefferson sent a letter to A. B. Woodward to thank him for sending his own classification scheme. He states that he now prefers Woodward's method of ascribing categories to types of science, not the faculties of the mind. This is because the scheme has a firmer basis in Matter, not in Mind, in the exterior rather than interior world of humans. (This is how the two most widely used schemes in the United States, the Dewey Decimal and the Library of Congress Classification systems, arrange collections.) For a long time Jefferson had viewed Jesus as a great moral teacher, but not God or the Son of God. Now he realizes the great moral teacher was a materialist as well, evidenced by the fact that Jesus laid great store in the resurrection of the physical body.

> Were I to recompose my tabular view of the sciences, I should certainly transpose a particular branch. The Naturalists, you know, distribute the history of Nature into 3 kingdoms or departments, Zoology, botany, mineralogy. Ideology or Mind however occupies so much space in the field of science, that we might perhaps erect it into a 4th kingdom or department. But inasmuch as it makes a part of the animal construction only, it would be more proper to subdivide zoology into physical and moral. The latter including ideology, ethics, and mental sciences generally, in my Catalogue, considering Ethics, as well as Religion, as supplements to law, and the government of man. I had placed them in that sequence. But certainly the faculty of thought belongs to animal history, is an important portion of it, and should there find it's [sic] place.

Jefferson saw the category of Memory or History encompassing all matters intrinsic to the natural world and the result of natural processes. Rocks, rivers, religions, and the use of the ablative case in Latin are understood alike as historical phenomena. Well before Darwin advanced his theory of how all life has a common ancestry and shared Nature, Jefferson, among many others, proposed that the category of Memory or History comprehends everything contingent, everything conditioned by time and place, everything construed as the product of historical forces. Philosophy itself was recast in Jefferson's conceptual catalog as simply another product of the human mind, itself contingent, historical, not universal and lawlike, like mathematics and the natural law.

In Jefferson's day, the word "materialism" was not yet tainted by fateful encounters with Karl Marx and the politics of dialectical materialism. Nor was it blighted by connotations of the banal love of luxury or nonessential material objects for their power to please, pamper, lend status, or provide psychological comfort. To identify the world completely with Nature deepened Jefferson's sense that the world is comprehensible through reason, even as he lamented that our own rationality—itself a by-product of history—could not be scientifically studied.

> Metaphysics have been incorporated with Ethics, and a little extension given to them. For, while some attention may be usefully bestowed on the operations of thought, prolonged investigations of a faculty unamenable to the test of our senses, is an expense of time too unprofitable to be worthy of indulgence.

Jefferson and his peers believed curiosity about the natural world could only lead to a comprehensive and coherent view of the planet, its inhabitants, systems, and principles, an idea developed by Jefferson's friend and proto-ecologist Alexander von Humboldt in his book *Kosmos*. They were confident that science would lead to a greater unity of knowledge, not to its splintering. It would foster a greater sense of well-being and integrity of spirit, not a sense of confusion, disorientation, distraction, and most certainly not a greater devotion to luxury.

"WHEN REASON FAILS, WE USE EXPERIENCE"

For Montaigne, experience was helpful if our reason could not find the answer. Today, in our empirical world, if we want to understand a problem, we look to experience—to evidence—first and then use reason to make sense of it. During Jefferson's lifetime the empirical perspective was embraced by naturalists who were digging up rocks studded with fossilized remnants of unrecognizable creatures. It was hard to make sense of the hard and fast evidence without calling into doubt the commonly assumed cosmological time frame. Understanding that rocks are clocks, geologists used fossils and rocks as demarcations of historical eras that could be dated relative to each other, if not absolutely.

Questions about the true antiquity of the Earth had been raised since the seventeenth century. But things came to a critical head when the true antiquity of humankind itself was debated in the context of growing consensus among geologists that time extended back hundreds of millions of years. Scottish geologist Charles Lyell (1797–1875) said as much in his *Principles of Geology*, published in the 1830s. In 1831, the young Charles Darwin took a volume of Lyell's book on his two-year-long voyage on the HMS *Beagle*. In his account of his travels, *The Voyage of the* Beagle, Darwin makes frequent reference to Lyell and other geologists as he tries to sort through various enigmas he encounters. What caused the extinction of native horses in South America? Why are there fossilized seashells in the Andean mountains? Everywhere he saw evidence piling up, slowly but inexorably, that the world came into being in a deep abyss of time, changes continuously, and that the laws that brought the world into being are active today, shaping the future. What is past has now become prologue. And if we learn how to read the past as it is written in stone, tree rings, ice cores, and fossils, we will learn about the future. It was the power of predictions inherent in materialism that captured the public and scientific minds alike.

That the past is prologue was a very old idea. Many Greeks had viewed matter as a form of memory encoding information not only about the past, but also about the future. The gods Proteus and Mnemosyne embodied the dynamic forces of matter and memory as divine beings, inhabiting both the mundane world of existence, accessible to our senses, and the extramundane world, indisputably real but beyond direct sensory perception. Among the first Europeans to recover the Greek view of matter as a dynamic elemental force that stores information was Francis Bacon, the powerful and influential advocate for empirical science. He was captivated by the story of Proteus, the god of the sea, whom Bacon writes about in *On the wisdom of the ancients* as "the Old Man of the Sea who never lies." He "plumbs the depths of the seas," and knows the truth of what is, what has been, and what is to come. But the Old Man of the Sea who knows everything communicates nothing—nothing, that is, unless one is able to capture him and hold him long enough to force him to speak. Here is the trick: Proteus is a shapeshifter and mutable, like water that can change from ice to vapor. He is as elusive to capture as the wind and as hard to hold as fire.

> "The addition in the fable that makes Proteus the prophet, who had the knowledge of things past, present, and future, excellently agrees with the nature of matter; as he who knows the properties, the changes, and the processes of matter, must, of necessity, understand effects and some of what it does, has done, or can do, though his knowledge extends not to all the parts and particulars thereof."

THE DISCOVERY OF DEEP TIME

While Bacon was interested in the implications of understanding matter's mutability, he did not pursue the question of the world's age. The geologists of the late eighteenth and early nineteenth centuries, though, found themselves backdating the age of the Earth in order to accommodate the evidence they uncovered. The lasting impact of their discovery of deep time—the millions and billions of years that elapsed before humankind and even life itself came along—is not a final and definitive birth date of the universe. That is still a matter of debate, under constant revision as more evidence about the infancy of the universe comes to light. No, the point is that it messed with a fixed and sacred chronology as such and changed the way we understand the fundamental processes of creation. Deep time is deeply generous to scientists. It allows for all sorts of random small alterations to accrete and add up to something altogether different—for change in quantity to scale up to a change in quality.

In hindsight we see Darwin's proposal that human beings evolved from primates as singularly traumatizing. Troubling as these ideas were (and continue to be for some), the truly momentous impact of materialism on the human psyche is what we have learned about the size, scale, and complexity of the material universe. As the universe got bigger, we got smaller.

William James pointed out that "the God whom science recognizes must be a God of universal laws exclusively, a God who does a wholesale business, not a retail business. He cannot accommodate his processes to the convenience of individuals." In exchange for the eminent powers over Nature we gained with the new science, we lost more than a god who knows our name and cares about our individual fate. We lost anyone to hold accountable for the ills of the world. We cannot blame our creator for the existence of evil. Nor do we have an all-powerful being to appeal to for help against our misfortunes. We do not blame God's wrath for plagues and droughts. Nor do we thank his benevolence for our health and good harvests. It also has serious implications for our collective memory, because materialism robs us of a supernatural source of knowledge. We do not believe in revealed truths and supernatural inspirations. This means that the cumulative knowledge and know-how of humanity, the collective memory that constitutes the entire repertoire of knowledge we need to understand ourselves and the world, is dependent on our careful stewardship. If we lose it, we cannot regain it through divine revelation. As we think about the future of memory in the digital age, it becomes clear that the stakes are very high indeed. Even though we may subjectively experience the world as overloaded with information, until we build the memory systems that will ensure the future access of digital information, it is all potentially at risk.

THE ORIGINS OF MATERIALISM

Where does this materialist view come from? Although materialism has had its proponents from earliest recorded times, the materialism that became the bedrock of contemporary science is the offspring of Christian theology, direct descendant of the belief that God is immanent in the world and that man, using the God-given faculty of reason, can read his presence in all of creation. Becoming adept in the language in which this knowledge is encoded is a sanctified endeavor. This was a view shared by many pioneers in the history of science—Roger Bacon, Galilei Galileo, Francis Bacon, Isaac Newton. As a consequence, the awe and wonder inspired by creation that belonged to the realm of religion was transferred to science—though God has been cut out of the picture.

Many of the scientists who established the principles and methods of materialist science were believers. They saw no conflict between religion and what they learned about the age of the Earth, because their interest was in *how* the Earth rose, the processes of creation and history. The questions of *who* and *why* were outside the scope of their inquiry. They followed the Galilean precept that scripture imparts moral law, but the Earth itself and all creation are the bible of God's natural laws. They saw no reason to put moral and natural law into conflict or competition for authority. It may seem a bit incongruous that the ability to build a case that this or that event happened in the past by assembling physical evidence could have the power to disarm religion and steal its magic—and even stranger that the church sanctioned and even encouraged this. But that is what happened.

It is certainly one of history's great ironies that religion—above all Roman Catholicism, but joined in time by Protestantism—

marginalized itself by encouraging the study of Nature as an act of religious devotion. The founders of the American Republic sought to protect the authority of religion by officially separating church and state. This was not intended to remove religion from public life, which is why we see the routine inclusion of prayers in Congress, the mention of God in federal and state oaths of office, and so forth. On the contrary, it was intended to allow a diversity of creeds to flourish and reduce the general undermining of religion by sectarian infighting. The Eastern Seaboard had, after all, been colonized by waves of religious dissidents from England and continental Europe. The disestablishment of the church meant that the pursuit of science and learning was protected behind a cordon sanitaire from sectarian battles. In a nation where there were many different and often competing religious persuasions, liberating the pursuit of knowledge from religious oversight seemed eminently practical as well as self-evidently moral. The particular role that learning played in the Western imagination first as a religious and then a civic virtue, and strengthened by the protections created in the United States to keep learning out of the cross-hairs of sectarian battles, set the stage for the rise of science, engineering, and information technologies in the century after Jefferson's death.

EVIDENCE AND THE FORENSIC IMAGINATION

The discovery of deep time created a forensic imagination, first and foremost among scientists, but soon among the general population as well. While forensics is associated in the public mind with legal cases and crime scenes, materialist science says that it is not just a crime scene that begs interpretation. It is the entire world. In this view, Nature is the ultimate archive, the most complete set of records about the past, the Universal Library itself. And science becomes the ultimate library card, granting unlimited access to the curious. Scientific investigation becomes a form of detection, and the key to solving the puzzle of "who, what, when, where, and how" is material evidence.

From the perspective of memory, the most consequential effect of embracing materialism is its unquenchable appetite for information in its smallest increments—single data points—and as many of them as possible. This was new. Certainly procuring evidence had played crucial roles in adjudicating disputes among people. The original cuneiform token was invented to be evidence, to bear witness as a visible and outward sign of a contract or agreement entered into by two or more parties and witnessed by a third. Evidence was used in Roman courts of law, where people were called together to judge a case based on information made public to one and all. The term "forensic" is used in reference to law courts and public testimony, and derives from *forensis*: of or before the forum where Romans appeared publicly for the disposition of criminal accusations. Evidence, in other words, is information available to all without prejudice. It is not esoteric, subjective, or privileged information. Even if not literally to be introduced at court, information has forensic value when it is reliable, publicly available, and authentic—that is, being what it purports to be, not a false representation.

The notion of judging truth by outward and visible signs did not die out with the Roman Empire only to be revived in the nineteenth century by scientists. Most trials in intervening centuries would introduce some form of evidence, such as the testimony of an individual with knowledge of the matter. Written records were deemed more reliable, because they were fixed and less likely to change than an individual's memories. But objective signs are not always written on paper. They can be written on the body as well. The seventeenth century, time of the glorious scientific revolution fought by Newton, Leibniz, Boyle, and Descartes, was also the time of witch trials throughout Europe and its colonies. And both pursuits of truth—of Nature and of the soul—placed great store in procuring evidence. During witch trials people would search for marks on a person's body to judge whether they were a witch or not, or they would thrust them underwater and wait for them to drown to prove they were innocent. The marks on a body or the lack of susceptibility to drowning were outward and visible signs of witchery. The point is not that we would not accept these signs as evidence, but that evidence is always something that a given group of people have equal access to in order to render judgment. And the criteria for judgment are known in advance and reflect a social consensus about truth. What counts for evidence is always culturally determined, as is its interpretation. This is as true today as it was in third-century Rome or seventeenth-century Salem. What distinguishes scientific use of evidence from other interpretive systems is the rejection of any and all supernatural and extramundane factors in either cause or effect.

By the 1830s, the growing demand for data of all sorts and about all things was well under way to becoming the data-intensive model of knowledge of today. Although there is no magic year or decade in which the current information landscape was sown with the seeds of digital data, we can see the 1830s as a juncture when all the elements necessary for a great information inflation were in place. We can call the 1830s the golden spike—the layer of earth geographers use to establish boundaries between geological time periods—to demarc-ate the boundary zone between an artisanal information landscape marked by scarcity and the technologically intensive forensic approach. Before the 1830s, only a handful of people saw the Earth as old and none saw humans as products of evolution. But in the 1830s we see—at least in the rearview mirror—two events that historians note as landmarks of the new knowledge landscape: the publication of Lyell's *Principles of Geology* and Darwin's assimilation of deep time into his theories about the origins of species. As J. W. Burrow points out, "Evidence of the antiquity of man in the conjunction of human artefacts and remains of extinct animals, which had been accumulating from excavations in France and England from the 1830s onwards, was finally accepted, after long skepticism, at the end of the 1850s."

THE INVENTION OF THE SCIENTIST

By the 1830s, the enterprise of science had begun its slow trans-formation from an amateur pursuit of knowledge, accessible largely to men of means, to a profession open to men and women from different ranks and economic classes. In this decade natural philosophy and natural history collapse to form the domain we now know (in English) as science, meaning the physical sciences. The merging of natural philosophy and natural history, together with the realignment of the natural sciences around evidence-based practices, was codified with the invention of the word "scientist" in 1833.

Though ubiquitous today, the word is new in the English-speaking world. It was coined by William Whewell, an eminent man of science who convened the third meeting of the British Association for the Advancement of Science. This group was formed to "promote science in every part of the empire" and was open to all "cultivators of science." After a stem-winding speech of welcome extolling the union of facts and theory, of natural history and philosophy, Whewell was publicly confronted by the renowned Samuel Taylor Coleridge. The man of letters strenuously objected to these humble practitioners with dirt on their hands and mud on their shoes (the stigmata of fieldwork) assuming the mantle of philosopher. Very well, Whewell amiably and shrewdly agreed. If philosopher is a term "too wide and lofty" for the likes of the assembled, then "by analogy with *artist*, we may form *scientist*" as one who is "a student of the knowledge of the material world collectively."

The word came only slowly into common parlance, not showing up often in written sources until the 1860s—and even then mostly in America. (For a long time, the English thought the word a vulgar American neologism.) Its invention nonetheless signaled that a scientist would be expected, like the artist, to have exquisitely keen powers of observation, to record one's observations faithfully, and to command a minimum of manual skills to craft tools of observation, to set up finely calibrated experiments, and to present the results with graphical as well as verbal precision. The term "scientist" grew in usage along with professionalization of the field. Though the pace and complexion of the change varied by country and class, by the 1870s the physical sciences were fully fledged as professions and supported by an educational system in most European and Anglo-American countries.

NEW MACHINES FOR NEW DATA

The forensic imagination demanded better instruments of investigation and lots of them. That desire to read Nature's archives drove—and still drives—the invention of new technologies to observe, measure, record, play back, analyze, compare, and synthesize information. The body of information spewing out of telescopes, microscopes, cameras, X-ray machines, lithographic printing presses, and calculating machines grew at a constantly accelerating pace. Libraries, museums, and archives were built with startling speed, filled up, added new storage units, and began to swell with books and scientific journals, maps and charts, photographs and sound recordings, natural specimens and fossils, cargo containers full of scientific booty from numerous expeditions to ever more remote corners of the world and beyond into space. The unified body of knowledge encompassed in Jefferson's library fractured into multiplying domains of specialized expertise that today can barely understand each other's methods and terminologies. The world of knowledge fissured, like the ancient landmass Pangaea that broke into seven continents. And like tectonic plates in constant motion, domains of knowledge collide and break into smaller areas of expertise, each supporting rapidly evolving, highly diversified forms of knowledge and producing more and more data every year.

Information technologies begin their dizzying cycles of innovation in media and data compression at this time. Before the 1830s, books were largely printed on linen and cotton rag paper and produced in small-scale printing shops. By the 1830s, the mechanical mass production of books using cheap wood pulp paper was under way. Before the 1830s the palette of recording technologies was limited to paper, ink, watercolors, and oils. Beginning in the 1830s, new technologies appear in rapid succession: image capture (the first daguerreotype was taken in 1839); sound recording (the first recording of a human voice was made in 1860 by Édouard-Léon Scott de Martinville and the first playback machine by Thomas Edison in 1877); and X-rays (discovered by Wilhelm Röntgen in 1895).

Before the Internet, distributing knowledge required transportation of physical objects. In 1830, the first intercity railroad, between Manchester and Liverpool, was built and augured the age of rapid overland routes to carry not only coal and grain but also books and magazines. But at the same time, immaterial modes of communication were under development as well. Electrical telegraphy was developed in several places in the 1830s (St. Petersburg, Göttingen, and London), and telephones were invented in the 1870s (Alexander Graham Bell).

The growth of new information technologies in turn created demand for a new information technology infrastructure—more libraries, archives, museums, collecting depots, specimen repositories, and skilled staff to manage all the assets they housed. By the 1830s, the volume of books to be collected grew to such a degree that these institutions became something quite different from the libraries that Montaigne or Jefferson knew. In 1836, the Library of Congress had 24,000 volumes, four times the number it had twenty years earlier. But it was a newcomer to the game. The British Museum (now the British Library) had 180,000 titles, the imperial library at St. Petersburg 300,000, the Vatican 400,000, and the royal library in Paris was nearing 500,000.

The proliferation of specialized libraries and archives in turn demanded a new architecture, designed specifically for purpose-built storage of collections at scale and reading rooms that could accommodate hundreds of bodies in chairs. The sheer quantity of information that was being assembled demanded specialization, and again, in the 1830s, libraries specialized in specific types of content for specific users. In 1832, the Library of Congress had accumulated so many books, manuscripts, and maps that its primary function—to serve the legislative needs of Congress—was overwhelmed by the sheer volume of the materials. Congress formalized a separate Law Library of Congress, moved it into its own space in the Capitol, and hired special staff to attend it.

The nineteenth century was marked by a series of crises around physical and intellectual control of all the evidence streaming in. Emboldened by the dream of accelerating the rate of human progress and well-being, expanding our control over the natural world, and freeing ourselves us from drudge labor, we went to work on documenting the world. We built more infrastructure to manage the documents, supported the growth of highly special-ized fields of knowledge to keep pace with the incoming data, and trained cadres of skilled professionals who in turn spontaneously differentiated themselves, like Darwin's finches developing differ-ently shaped beaks. Technologies and tools coevolve with the ideas and cultural practices of their users. And then the users outgrow them and want more. Science itself cannot advance without ever finer tools of observation, measurement, experimentation, and the communication of these results to other scholars. (The World Wide Web was devised to speed communication of research results among collaborating scientists at distant sites.) Thus the happy relay race between demand for and supply of new technologies to observe, measure, and record accelerates quickly into the giddy pace of the Red Queen and Alice running faster and faster just to keep up.

THE ART OF DEDUCTION

The art of forensics is in seeing matter as the footprint of the past. This is an aesthetic intuition based on the recognition of certain patterns that recur in Nature and are often described by scientists and mathematicians as elegant, beautiful, or simple. The science of forensics lies in the patient application of refined skills to decipher the clues left by previous states of existence. It takes decades of education to develop the scientific sensibility to detect significant patterns and acquire skills to decode the message in the matter. But the premise itself is breathtakingly simple, something anyone can grasp intuitively, without any specialized knowledge. The aesthetic appeal and metaphysical profundity of the forensic insight are the reasons the whodunit-and-how fiction of detection arose in the nineteenth century and still reigns as the grand narrative of our technological age, the Age of Matter.

By the 1840s the forensic imagination had already penetrated popular culture. In 1841, Edgar Allan Poe published the first story of detection, a "tale of ratiocination." "The Murders in the Rue Morgue" features Monsieur C. Auguste Dupin, a man of science, who uses the scantest of physical evidence to arrive at an improbable yet true deduction about the murderer of two women—an orang-utan. Arthur Conan Doyle acknowledged Dupin as an inspiration for Sherlock Holmes. Beyond the intrinsic charm of an eccentric detective, what makes Dupin, Holmes, and legions of their descendants compelling to watch is the near mystical marriage of knowledge and practice to interpret the world put in the service of moral actions—the catching of criminals and transgressors. Their zealous, even ascetic, dedication and single-mindedness of purpose became the hallmark of the professional man (and eventually woman). Immersed in the data-dense environment of a crime scene, Sherlock Holmes always brought laser-like focus and purpose. He believed that being a specialist meant that he had to pick and choose what he paid attention to. "You say that we go round the sun," he said to Watson in one of his periodic fits of pique at the obtuseness of his friend. "If we went round the moon it would not make a penny-worth of difference to me or to my work."

The main reason materialist science triumphed over all competing models of knowledge lies in its effectiveness in prediction. By eliding a deity who can intervene at will to suspend natural laws, we have gained considerable control over our own destiny. Science gains its powers of explanation by conscientiously separating "the objects of natural knowledge from the objects of moral discourse." Scientists separate *how* questions from *why* and dwell exclusively on *what is*, not *what ought to be*. This is the moral hazard Socrates warned against—that by alienating our knowledge, making it "external to us," we have bought an immense measure of power over the world at the expense of having power over ourselves. The nineteenth century saw the rapid rise of specialization, and not only in domains of knowledge. Labor, too, grew increasingly specialized as factory production adopted the assembly line. When Karl Marx described the "alienation of labor," he was not just talking economic theory, but also of the increasing perception that laborers were losing a sense of autonomy. As we outsource more of the most intimate part of ourselves—our personal memory and identity—to computer code, the fear of losing our autonomy—the alienation of our data, so to speak—increases because in the digital age, only machines can read our memory and know what we know at scale. As we gain mastery over our machines, this anxiety will lessen. But it will never go away, for the trade-offs we make between our appetite for more knowledge and our need for autonomy and control will continue to keep us on the alert for unintended consequences.

Science can create very powerful technologies, but it is not science alone that can help us manage them. As the historian Steven Shapin says, "The most powerful storehouse of value in our modern culture is the body of [scientific] knowledge we consider to have least to do with the discourse of moral value." Today we ascribe virtue to those who advance the progress of humankind rather than the salvation of souls. The elite vanguard of virtue are no longer priests and clerics dressed up in black and crimson robes, but rather scientists, engineers, and technology entrepreneurs dressed down in white lab coats and denim jeans. The tale of progress from ignorance to enlightenment follows the same plot-line as the narrative from sin to salvation. This is a distinctly Western perspective, quite different from the cyclical view of time typical of Hindu and Buddhist world views. But it is the one that undergirds the global language of digital memory.

The conversion to materialism was the critical change in consciousness that led to our current dominion over the world. The cluster of causes that reinforced the forensic shift is not hard to isolate: the embrace of empirical methods and materialist theories to understand natural causes and effects; the harnessing of that understanding by economic systems that apply this knowledge to create ever finer instruments and accumulate greater amounts of evidence; the dedication of resources to educating an expert work-force to generate and apply more knowledge; and political regimes that keep the pursuit of scientific knowledge well resourced and safe from either religious or ideological interference. Take any of these four factors away, and science and technology are crippled. But taken together, these forces create a runaway effect that results in an inflation of information. Such an outcome was probably inconceivable to Thomas Jefferson, and no aspect of it would be more confounding than its effect on his library.

CHAPTER SEVEN

THE SCIENCE OF MEMORY AND THE ART OF FORGETTING

What I perceive are not the crude and ambiguous cues that impinge from the outside world onto my eyes and my ears and my fingers. I perceive something much richer—a picture that combines all these crude signals with a wealth of past experience . . . Our perception of the world is a fantasy that coincides with reality.

—CHRIS FRITH, COGNITIVE PSYCHOLOGIST

LIFE WITH NO LIMITS

IN THE NINETEENTH CENTURY, we renounced intellectual limits, a priori censors, and religious filters on what we allowed ourselves to know. Emboldened by the Enlightenment cult of reason, we saw curiosity no longer as a vice, but as a civic virtue. We institutionalized public support of libraries, schools, and a free press to propagate progress and freedom. At the end of the twentieth century, the invention of digital computers vaulted us over the physical barrier to ideas spreading like fire "from one to another over the globe." Information can be shared ubiquitously and nearly instantaneously.

Now we confront a new barrier—the natural limits of human attention and ability to absorb information. These are biological constraints shared by all living creatures. The analog memory systems we have superseded were well adapted to this limitation, constraining the production of and access to print and audiovisual content. To be equally effective, the digital memory systems we build will need to acknowledge and compensate for our natural limits. The biology of memory is a young science, still in its salad days. Yet even preliminary findings suggest how to work with, rather than try to defeat, our brain's architecture of constraints to enhance personal and collective memory. Memory sets up a series of gates and controls to turn away the trivial or distracting and allow ready admittance of valuable information for conversion to long-term memory.

Scientists learn about how memory works in large part by studying what happens when memory breaks down. Two types of memory failure have particular implications for memory in the digital age. The first failure is the interruption of long-term memory formation. Without conversion of short-term memory to long, we are not able to make sense of the world, recognize patterns, make inferences or conjectures; in short, we cannot learn. Without long-term memory, we would be stuck in the present. Every day would be Groundhog Day, only with no happy Hollywood ending in sight. The second failure is the loss or disintegration of memory. Amnesia robs us of the ability not only to remember the past, but also to imagine the future, make predictions, and engage in mental time travel. If we lose the long-term memory of humanity, we will be like amnesiacs, not knowing who we are, where we have been, or where we are going.

Biology is neither personal nor cultural destiny. But as we explore the early-stage findings of memory science, we discover that the relationship between the part and the whole, the individual and society, personal and collective memory, are intimately intertwined and evolve together. We may never be able to map cultural history directly onto our biological legacy, but we cannot afford to ignore it. To the extent that our cultural life extends and elaborates on our essential biological capacities, what we learn may serve as a guide, if not a set of detailed instructions, on how to design digital memory.

HOW WE LEARN THE WORLD

The crowning achievement of memory is the model of the world
that we carry in our heads, a kind of diorama that closely resembles
reality and allows us to respond to events in real time. As Frith
points out, what we perceive in any moment is a combination of
real-time perception and stored information, our memory of the
world. No creature is able to process enough information in real
time to react appropriately to events as they transpire. So it prepares
a miniature map of its environs and annotates each spot with
detailed information about people, places, and things found there.
Like the sea slugs and rats studied in labs, we are all born cartog-
raphers who draw mental maps to orient ourselves to our surroundings. In split seconds, the mind runs through its geo-checklist:
Where am I? How did I get here? Where am I going? Who else is here?
Like Simonides, we spatialize that information, sometimes in way-
finding maps that tell us how to get from point to point, sometimes creating a *mappa mundi*, a map of the world that represents
our conception of space and time. Humans are also aware of
existing in the fourth dimension of time. We create narratives that
make sense of our passage through time just as our maps make
sense of our passage through space. The model we assemble has to
be flexible and easily modified over time so that our behaviors can
be calibrated to the current scenario.

How our brains decide what information is important to
remember is hidden from our direct observation. Most of the
processes of attention and retention of information are directed
by our instincts or emotions and processed unconsciously. As neu-
roscientists Eric Kandel and Larry Squire note, "By virtue of the
unconscious status of these forms of memory, they create some of
the mystery of human experience. For here lie the dispositions,
habits, attitudes, and preferences that are inaccessible to conscious
recollection, yet are shaped by past events, influence our behavior
and our mental life, and are a fundamental part of who we are."
The unconscious nature of these nondeclarative memories makes
them essentially off-limits to analysis by machine intelligence, at

least at our present state of knowledge. Unlike their facility with logical processing, computers and robots do not process information with emotions. Nor do they store emotional memories. As we will see, emotions are essential to decisions involving values, such as making choices between leaving work at the end of the day to spend time with family or staying late to finish a pressing assignment, and between telling a friend a white lie or seriously hurting his feelings.

We are not born as blank slates, having to learn literally everything from scratch. Living creatures come with preprogrammed memory, the genome, that encodes the history of the species and provides full instructions on how to become an ant if you are born with ant genes, a marmot if with marmot genes, a human if with human genes. Scientists will often refer to the genome with rhetorical flourish as the book of life and modern biology as a science of information because so much of their work lies in deciphering how cells communicate with other cells and send orders from one part of the body to another that are instantly heeded. Because genes retain modifications over time, scientists identify the genome as an archive of the past. "The new genome in a fertilized egg is not the new person nor even an encoded version of the person; it is the archive of information that the developmental clock will use to form a new and genetically unique body from the descendants of one fertilized egg cell." In other words, the egg is like an executable file: Add food and water and watch it grow. We are programmed to be curious and acquire information about the specific environment into which we are born. The more complex the environment, the steeper the learning curve and longer the education we experience. Social animals like *Homo sapiens* and elephants require long gestation periods and decades of on-the-job training before they reach maturity.

Conceptually, memory formation and retrieval is relatively straightforward, the very model of curation. It boils down to just a few steps: selection, acquisition, categorization, storage, and preparation for ready retrieval on demand. That said, each step of mental curation is intricately detailed and involves global coordination and synchronization with other processes in the brain. How this happens is beyond the horizon of present-day science. But the basic contours of each process are coming into focus.

We scan the environment for information that catches our attention. Given that the brain's primary job is to keep us alive, it is highly attuned to perception of novelty. We become inured— habituated, as scientists say—to the familiar. Something new and unexpected will grab our perceptual attention, and the brain will make inferences about everything else, based on what it has laid down in its memory. As a consequence, most of the information we perceive in any given moment is disregarded—effectively thrown away—because it is redundant. What we acquire through our senses is instantaneously processed through emotional and cognitive centers for value and sensemaking. The brain looks for matches against similar information stored in the brain, and then temporarily parks it for ready reuse. Short-term memory is packaged first as read-only, as it were, and is easily discarded if not transferred into long-term storage. At the end of the day, everything taken in gets parked in the overnight parking lot, so that it can be sorted and processed during sleep before going into its assigned address. The crucial processing steps that occur during our sleep are inaccessible to our consciousness. By day, our strategy for sensemaking is to create a narrative of events that suggests cause and effect and identifies a context in which all the elements cohere and find meaning. By night, the linear timelines we observe during waking hours are not operative. In dreams the rules of cause and effect are suspended, our internal censors fast asleep and vast realms of reality normally inaccessible to us take center stage. In sleep the mind sorts the rubbish of the day from the riches we grab and keep forever. Adrienne Rich said poems are "like dreams: in them you put what you don't know you know." But the reverse is true as well. Dreams are like poems: In them what we know is represented by symbols whose meanings are ambiguous, multivalent, open-ended.

Information has value to the extent that it *has the potential for reuse.* Erwin Schrödinger noted: "Biological value lies only in learning the suitable reaction to a situation that offers itself again and again, in many cases periodically, and always requires the same response if the organism is to hold its ground." What is useful gets converted to long-term memory through a consolidation process before taking up permanent residence in our mind. During consolidation, content is abstracted from its native context and made available for reuse in other contexts. After once experiencing the scorching heat of an open flame on flesh, we abstract that pain into an infinitely reusable lesson about fire, and further, all forms of intense heat and, metaphorically, about pain associated with emotional intensity.

Information processing may occur in very specific parts of the brain, but the storage itself is likely to be organized and distributed in a way that facilitates an almost infinite number of possible uses in a wide variety of scenarios. Our uncanny ability to see patterns everywhere allows us to interpolate information from our memory banks into present perceptions. What we "see" are inferences the brain makes on the basis of what it already knows—that is, remembers. The patterns we see are based on samples of the world, the data within each sample are chemically tagged with specific molecules that express value, and those with the greatest salience command our attention. You may be walking down a strange street in a foreign town when out of the corner of your eye, you see a fast-moving blur approaching you. Before you have time to look properly and assess what it is, your mind has already commanded your body to protect itself against a collision by stepping back. Your mind doesn't know what it is, but it has already deferred to memory and taken evasive action. These mental patterns, whether of self-protection or pursuit of pleasure, are like the flexible yet sturdy and resilient lattices of a pergola, serving as architectural supports for the buds of thought that sprawl like vines. This is how we wind up seeing a complete picture of the world from split second to split second, even when our senses have time only to record fragments of what we perceive in any given moment. Like poetry, mathematics, and music, good memory relies on patterns that simultaneously constrain content and suggest meaning.

During recollection, a memory is opened up like a book or computer file, gets reworked, then reencoded and stored in a slightly modified way. "The retrieval of the consolidated memory is a dynamic and active process in which remodeling or reorganization of the already-formed memories occurs to incorporate new information." Recall is literally a rebuilding process, executed chemically, and new perceptions are incorporated into the old. The more often a memory is called up for use, the stronger it gets, be it declarative (factual) memory such as an event or word or somatic (physical) memory such as a smell, a sound, a golf swing or keyboard skills. Memory consolidation, "the progressive post-encoding stabilization of the memory trace," is the phase in which connections between the new data coming into the brain find their home in existing mental contexts or webs of associations. In other words, memory consolidation creates meaning by putting information into an appropriate context. Once in context, it can be used again in the future. And it is in this phase that a memory is most vulnerable to losing its way, never finding a meaning or context that can hold it stable and available for future use. Because every memory is tweaked and fortified in use, we live in a state of nonstop historical reinterpretation. The upshot is that the past itself changes in the process of remembering. The representations of things that were laid down previously—whether five decades ago or five minutes—are modified simply by being used. In use, they are placed within the environment of the present moment and that use becomes an intrinsic part of memory itself—the *past plus*.

Before the twenty-first century, we relied on a system of preserving memories that complemented our internal memory. One of the breathtakingly simple advantages of the cuneiform, scroll, or printed page was that the memories inscribed on them were not easily changed, overwritten, or erased. On the contrary, these durable objects acted in exactly the opposite way our brains work. If kept in reasonably good physical shape, the words and images on a piece of paper would not change one whit for hundreds of years, no matter how many times they were read. Digital memory operates much more like biological memory. It is not really fixed and is easily overwritten or updated without leaving much trace of the changes made. With digital memory we lose one of the crucial advantages of fixed and stable physical memory—the fixed and passive retention of information. How we re-create the advantages of physical storage in the digital realm is an important consideration in designing our memory systems to stabilize digital data over long periods.

Deep learning and creativity, on the other hand, rely on the transformation of one day's intake of perceptions to something sustained over time, embedded within a network of existing associations. What we call creativity is the use of mental content in contexts wildly unrelated to its source. (Humor is putting the content into wildly incongruous contexts.) But this deep absorption of content into the mind can exact a price. One common byproduct of preparing content for multiple uses is source amnesia, whereby we remember content strongly but not where we first acquired it. You may know a certain film was recommended to you, but you cannot remember by whom. You may have read how many people were killed in an accident, but not where you read it. In each case the source was familiar or trustworthy enough (your friend who recommended the film) that you did not bother to store information about it, or the information so arresting (the fatalities) that the content eclipses all other impressions.

ANALOG WAVES AND DIGITAL BITS

The brain uses both analog and digital processing. Digital signals communicate what to pay attention to, and analog processing comes in to suppress some signals and amplify others in order to allow us to focus our attention on the object of choice. This is how, for example, we can catch the voice of the person we are looking for in a noisy room. One process tells the brain where our friend's voice is coming from, and another part dials down the volume of other voices and dials up the volume of our friend.

Analog circuits process continuous variations in intensity; digital circuits operate in an on–off/yes–no mode. So far, we have not been able to devise machines that use both analog and digital signal processing with anything like the specificity and flexibility that our hybrid brains do. For most of our history, we have relied on analog processing—printing ink on paper, painting pigments on canvas, inscribing sound waves onto wax cylinders or plastic discs. We are long familiar with the strengths and weaknesses of analog formats. They can be slow and imprecise, but infinitely rich in subtlety and nuance of perception.

By its inherent nature, digital recording samples information, cuts it up, disambiguates it into 0s and 1s, and packages these bits in a way that makes them very easy to manipulate, reorder, and, most powerfully, send over long distances with minimal loss of information. This creates a paradox: that analog information has a greater integrity—of the literal kind—and more accurately mirrors the embodied mind in time and space. But it is less flexible and faithful as a means of analysis, comparison, and communication over long distances. The constraints of a book, a photograph, a paper map are precisely those of the human body—unable to leap tall buildings in a single bound or teleport itself over long distances in seconds. As man-made physical objects, all these artifacts of recorded knowledge—maps and photos, books and magazines—exist on the same scale as the humans who created them. The digital does not.

With digital encoding comes a fateful dependence on machines for reading and playback equipment. A digital file is machine readable, not eye legible. We can make little or no sense of it unless we have the right hardware to run the right software to decipher it. A dependence on playback equipment is not new to digital information. It began with sound recordings. It takes only ambient light to read a musical score but any and all media that carry sound waves, from wax cylinders and wire to lacquer discs and cassette tapes, demand machines to convert grooves and magnetic signals into sound waves. The introduction of recording technologies that depend on machines to read them, dating back to Edison's invention of the phonograph in 1877, marked the beginning of our unconditional dependence on secure and reliable sources of power to maintain our knowledge and memory banks.

Digital moves us in great leaps and bounds to new efficiency and flexibility of the types of information we can record. And it makes manipulation of what is recorded easy, because everything is already cut up into little bits that can be shuffled and rearranged, tweaked for saturation of color or audio wavelength. But it does so with corresponding losses in our ability to distinguish amplitudes, continuities, and intensities. What the demotion of the analog capacity for detecting and capturing degree and continuity means for our culture, for ourselves, is unknown. But we know the effect of accelerated processing time and of binary thinking in our everyday lives: We have simultaneously more information and fewer means to sort its value.

Digital recording technologies change the way a musician approaches performance; technical proficiency (or deficiency) is more audible in digital recording than degrees of expressiveness. The difference between capturing light on photochemical film and on digital memory chip has similar effect—greater sharpness, less subtle gradation of palette. At the extreme, digital can pixilate and reveal its complete inability to represent the nature of reality as we experience it—continuous and deeply connected. As the cognitive scientist Randall O'Reilly says, "It is clear that the brain is much more like a social network than a digital computer." Memory and learning are investigated now as products of "the graded, analog, distributed character of the brain." It turns out that the computer is not an accurate metaphor for the brain.

VALUE AND EMOTION

Emotion is the body's internal representation of value. Emotions such as fear and joy register the meaning of something to us. Their explicit and conscious counterparts—anxiety and happiness—are what we say we feel. Are we gripped with fear when weighing our

chances of doing well on a crucial test we do not feel prepared for?
Then thinking of it makes us anxious. Are we suffused with joy
when we spend time with our children after a long absence? Then
thinking about them makes us happy. Emotional value is processed
preconsciously and tagged by chemical markers. The stronger the
emotion, good or bad, the greater the value for better or for worse.
Sometimes the emotions are passed to the conscious mind and we
become aware of them, sometimes not. But whether conscious or
not, they are associated with our senses through visual, olfactory,
or aural cues. As the neuroscientist Susan Greenfield writes, "Pure
emotion can be viewed as the core of our mental states . . . as when
we are infants, feeling is not greatly tempered with individual
memories, with cultural or private meaning, or, most important of
all, with the self. Feelings just are." No matter how long we live
and how polished our manners, emotions can never be civilized. If
they were, they would lose their value to us. They are meant to
surprise us.

Because emotions are core to assessing value, the healthy brain
does not waste space by storing data raw. We always cook data, and
the most important information is marinated in heavy solutions of
emotional cues. As we process perceptions, we extract the most
significant and map them into networks crisscrossing the nervous
system. Most of what we learn bypasses our awareness altogether
and goes straight to the emotional and instinctual centers of the
brain. This is for our own safety. Experience modifies behavior
"without requiring any conscious memory content or even the
experience that memory is being used." Directed attention requires
effort and takes time we do not have when we are out and about in
the world. As noted, our unconscious attention is invariably drawn
to what is new, unfamiliar, bright and shiny; or what is in motion,
shadowy and sinister; or what triggers some emotional shudder.
Without our being aware of it, attention is caught in the half-
second it takes for a perception to come into our consciousness. By
then we have already "made up our mind" about what we perceive
and how to react to it. We can slow down to consider and change
our minds. But we seldom do. Our split-second judgments often

pass for conscious decisions, but that is an illusion. We like to say that memory plays tricks on us, and often it will present the past differently from what we actually experienced at the time. But this is a reminder that memory has many jobs other than keeping track of the facts as they occur.

Emotions are supremely significant because of their determinative role in belief and decision making. As the neuroscientist R. J. Dolan writes, "Emotion exerts a powerful influence on reason and, in ways neither understood nor systematically researched, contributes to the fixation of belief" as well as memory. From the time of the Enlightenment onward, Western culture has deemed reason a stronger and more prestigious form of intelligence than emotion. Reason thinks slowly, not intuitively, and is effective as an instrument used to exert human will. But it is not the font of empathy and fellow feeling that social life requires. Computers, on the other hand, can reason with stunning speed, but they cannot simulate human decision-making processes with equal speed because they are not emotional. They can learn to simulate our behaviors by assessing the outcomes of past choices to make probabilistic predictions ("people who liked this also liked . . ."), and often that is good enough.

Reason is a rarefied creature, demanding concentration and concerted blocks of time. Consequently, we use it sparingly. We rarely expend the time and energy to sort out something unfamiliar, let alone something that creates dissonance with what we already "know." Information that fits into preexisting categories and values is simple to incorporate. On the other hand, it is clearly self-defeating if acquiring new—let alone contradictory—information is really hard. Evaluating what we perceive depends on deep processing that operates over long periods of time and can be neither rushed nor short-circuited. This explains why a tired brain is less likely to acquire new information than one that is well rested. It also explains what is known as *confirmation bias*, whereby we pay more attention to information that confirms our existing view of how the world works than to information that contradicts or complicates that mental model. This is the paradox we live with: The mind's struggle to accommodate more and more information in our inelastic brain can end up making us less open to new information, not more. To a large degree, wisdom is a feature we can acquire over time to compensate for this, developing a heightened ability to assess how valuable any new piece of information may be in the larger context of what we already know.

DISTRACTION: THE STORY OF S.

A man whose brain was far too elastic for his own good was the Russian known to us as S. His story is told with compassion and discretion by the psychologist Aleksandr Romanovich Luria in *The Mind of a Mnemonist: A Little Book about a Vast Memory.* (Luria refers to this man simply as S., and we shall observe his discretion.) Luria studied his subject over the first three decades after the Russian Revolution. S.'s problem was that he did not have what Luria calls "the art of forgetting," a faculty of the healthy mind that creates order from excess by converting selected short-term memories into long-term memories and flushing away the rest. Long-term memories in turn align themselves into distinct patterns as they are abstracted into general categories and reused over time. S. remembered a prodigious amount of detailed information that never lost its vividness, but at the high price of never abstracting perceptions into greater patterns of significance.

As a consequence, S. suffered from a *disorder of distraction*: He could not make things dull, and had a hard time maintaining focus on anything for extended periods. He was unable to sort his impressions for value and emotional salience. To him the world was far too vivid far too much of the time. Luria reports his patient could remember everything but was unable to establish priorities for his memories. As a consequence, his speech was digressive and prolix. He would start out on one subject and end up somewhere very far away, often down a blind alley. He easily confused what he had remembered (because everything he encountered in his

Truths sacred and undeniable, or merely self-evident?: The Rough Draft of the Declaration of Independence, drawn up by Thomas Jefferson and corrected by Benjamin Franklin and John Adams.

Paleolithic selfie: About 30,000 years ago humans stenciled their handprints with red ochre in the Chauvet Cave of southern France.

" 'Lord save us!' cried the duck. 'How does it make up its mind?' "

Of two minds: Pushmi-pullyu from Hugh Lofting's *Doctor Dolittle*, blessed with two heads but unable to make progress or walk a straight line when they disagreed.

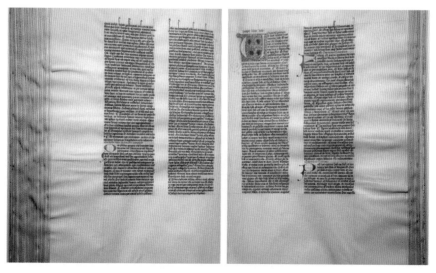

Reading the gods' intentions: Seventh-century B.C. cuneiform tablet providing expert interpretation of human and animal birth defects as divine omens.

New technologies imitating old: Manuscript Giant Bible of Mainz, left, and print Gutenberg Bible, right, meant to look handmade.

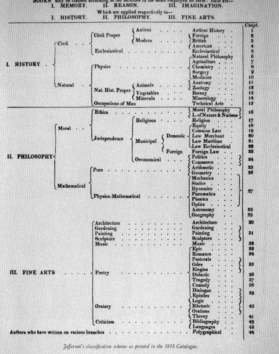

Jefferson's classification scheme as printed in the 1815 Catalogue.

The faculties of Thomas Jefferson's mind: Catalog of Jefferson's 6,487 books, drawn up in 1815 when he sold them to Congress. Two hundred years later, the Library of Congress has over 160 million items in all media.

Reading as enlightenment: Sainte-Geneviève Library in Paris (1838–1850). To allow natural light and discourage fires, Henri Labrouste used newfangled cast-iron for the reading room which could seat hundreds of students in search of knowledge.

Memory as meaning: Michel de Montaigne, early print native, invented the essay as a tool of self-knowledge. He continuously revised his essays as he changed over time.

Disorder of distraction: Russian psychologist A. R. Luria, whose subject S. remembered everything, lacked the "art of forgetting," and consequently felt his life "took place in dreams, not in reality."

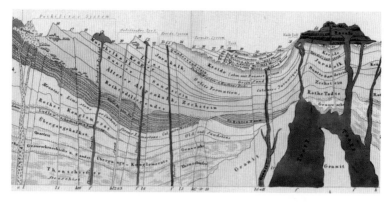

A slice of time: This detail of a geological map from 1841 shows the earth shaped by dynamic, violent forces over millions of years, made vivid by hand-painting over engraving.

Making pictures of sound: Experimental physicist Carl Haber, whose work on devices to image subatomic particles is the basis for technologies that rescue sound from recording media too fragile to play.

Big Data: Corridor of the Library of Congress in 1897 during the move out of the Capitol into a magnificent new home the size of a city block. The library outgrew its new building within a decade.

What digital memory looks like: Brewster Kahle, in the Internet Archive's stacks of machine-readable digital memory, where books, audio, and video all live on the same drive.

daily life triggered a chain of recollections) with what had actually transpired. Memories were so fresh in affect and spun out in his mind so rapidly that he mistook his recollections for reality. There were periods in his youth when he did not get up in the morning to go to school because even thinking about arising stimulated memories of having done so before. He thought that he had gone to school even as he lay still under the covers. After some period of time, S.'s inability to distinguish between what he had recollected and what had actually transpired led to a blurred sense of reality, an attenuated sense of actually being alive.

The proximate cause of his overarticulated memory was likely his synesthesia, an enigmatic neurological phenomenon whereby stimulation of one sense or perception provokes another unrelated one. When he recalled the word for beetle, *zhuk*, for example, he immediately thought of "a dented piece in the potty, a piece of rye bread," and the entire sensation of turning the light on in the evening and having only a part of the room—the *zhuk*—in light. Once he thought of these, he could not let go of them. On the one hand, it was easy for S. to establish a series of associations that would prompt recall. He was able to make a living as a mnemonist, committing vast amounts of detailed information to his memory and recalling them on demand in a performance. On the other hand, the sheer density of sensory associations inadvertently set off by the slightest of provocations made everyday living distressing. Whatever other gifts he may or may not have had, synesthesia exacerbated the effect. Associations among senses effectively superseded any association among events and disconnected him from an ordinary sense of time passing. The net effect, Luria noted, was that S. lived dispossessed of the present, always in expectation of great things about to happen.

His affliction was destined to get worse with time as he accumulated more and more information and had trouble ridding himself of any of it. To compensate for the fact that he was forming memories in an uncontrollable way, he developed a technique to surround a specific memory with as little context as possible. He would narrow the frame of reference of a memory, tightening the context like a noose around a given fact he wished to recall, thus strangling all possible associations. How could this not bring about the effect of making the connection between things opaque, invisible, and dissociative? As a consequence, his sense of narrative broke down. When S. processed new information, everything went into Save and nothing into Delete because it was redundant, or merged into another file of similar memories. He could not consolidate the content of his memories, to "convert encounters with the particular into instances of the general." The information he took in could not be compressed, abstracted, or generalized through pruning a memory of its redundant or irrelevant data points.

A telltale sign of his disability was his difficulty in reading. Every idea or concept called forth a torrent of imagery, most of it extremely vivid and tangential. "This makes for a tremendous amount of conflict and it becomes difficult for me to read. I'm slowed down, my attention is distracted, and I can't get the important ideas in a passage." Understanding poetry or in fact any figurative use of language was well-nigh impossible. He could not grasp synonyms or homonyms used in different contexts. He was perplexed by the use of the word "arm," for example, to describe something both attached to a person and extending from an institution—the arm of his wife, say, versus the arm of the law. An expression such as "to weigh one's words" confounded him. Poetry, by its very nature graphic and figurative, was beyond his comprehension because he could only understand the images literally. Paradoxically, poetic images so powerfully stimulated his own mental imagery that they could not evoke anything other than the very first meaning that came to mind, which was the literal meaning.

More serious were the implications of his inability to filter out distractions for his ability to follow a narrative. He could hardly track anything that changed over time, and that, according to Luria, was why he had a hard time remembering faces. He could not pick out any prominent or distinguishing features in a face because the face changed and was always so expressive. To him, every face was a narrative he could not follow.

The model of the world that S. carried in his head was never made coherent and continuous in the flow of time. It was simply flooded with more unassimilated data. He was incapable of making plans because he actually had no sense of how things happened. It was just one damned thing after another. "At one point I studied the stock market, and when I showed that I had a good memory for prices on the exchange, I became a broker. But it was just something I did for a while to make a living. As for real life—that's something else again. It all took place in dreams, not in reality."

S. was painfully aware of his problem. "I was passive for the most part, didn't understand that time was moving on," he said. "All the jobs I had were simply work I was doing 'in the meantime.' The feeling I had was: 'I'm only twenty-five, only thirty— I've got my whole life ahead of me.' . . . But even now I realize time's passing and that I might have accomplished a great deal— but I don't work. That's the way I've always been." Luria poignantly adds that "he had a family—a fine wife and a son who was a success—but this, too, he perceived as though through a haze. Indeed, one would be hard put to say which was more real for him: the world of imagination in which he lived, or the world of reality in which he was but a temporary guest." S. felt there was some great and consequential drama transpiring somewhere, some great river of time and meaning that was nearby and he was sure to find at some point in time. But at what juncture in his life could he push off from shore and enter the mighty, relentlessly moving river of life? At the end of his life, S. felt detached from his own life, as if he had never really lived it.

DISTRACTION AND DISRUPTION

It is easy to read S.'s life story as a cautionary tale about the temptation to save all data because our capacity for digital storage keeps growing. The quantity of data we amass does not by itself add up to a grand narrative. Good memory is not data storage. Forgetting is necessary for true memory.

We have created a technologically advanced world that operates
at a breathless pace, driven by a culture that demands more innovation ("disruption"). But science tells us that this disruptive and
accelerated pace is self-defeating because our bodies and minds still
operate at the same tempo as they did in ancient Sumer. Analog
memory systems based on objects had a built-in friction that slowed
us down and demanded depth of focus and concentration. Digital
gets rid of all the friction, speeds things up, and taxes our powers of
concentration and discrimination.

We are a culture obsessed with facts—the intrinsic value of a
fact. But our brains do not share this reverence for facts. The
human brain plays fast and loose with them. For facts are cultural
not natural phenomena. What the mind seeks is meaning: a sense
of order, of appropriateness, of measure and purpose. To that end,
the brain finds impressions most useful for its business. From
impressions, what Frith calls "the crude and ambiguous cues that
impinge from the outside world onto my eyes and my ears and my
fingers," the mind will create a very faithful diorama of the external
world, but it does not correspond to a factual representation.
We have to interpret facts and impressions in the context of our
environment in order to make sense of them. Facts have no intrinsic
meaning. It is culture that creates expectations of what makes sense
and what does not. If all the mind does is create literal representations of the external world, then all cultures in all times and all
places would have exactly the same working models of the world.
They do not. Each culture spawns its own interpretive framework
by which people make sense of "the facts on the ground."

The architect Le Corbusier (1887–1965), who self-confidently
placed himself at the very vanguard of innovation, wrote that
there are "living pasts and dead pasts. Some pasts are the liveliest
instigators of the present and the best springboards into the future."
In the twenty-first century, as we live longer and the pace of
change accelerates, our mental models of the world need to be
increasingly flexible and easily updatable. Yet the faster things
move, the harder it will be to maintain a strong sense of continuity. In periods of information inflation such as ours, it is all too

easy to feel like S., constantly assailed by vivid images we have no time to digest, sort through, and disregard and discard if not of lasting value.

Le Corbusier had it right: The past is a plural noun. How do we distinguish between living pasts and dead, between what actually happened and a faux history, full of false facts and narrow windows into the future? With the pace of change accelerating, do we even need to bother with the past? If Thomas Jefferson, ardent revolutionary and futurist, is right, the more we care about the future, the more we need a rich, diverse, accessible record of the past. Because memory is not about the past, it is about the future.

Chapter Eight

Imagination: Memory in the Future Tense

Time without consciousness—lower animal world; time with consciousness—man; consciousness without time—some still higher state.
—VLADIMIR NABOKOV, *STRONG OPINIONS*

IN 2011, A TEAM OF scientists published a research report about "How to Grow a Mind: Statistics, Structure, and Abstraction." They wanted to know how our minds build a world far richer than anything it knows from its own experience. Couched in the language of information processing, they asked:

How do our minds get so much with so little? We build rich causal models, make strong generalizations, and construct powerful abstractions, where the input data are sparse, noisy, and ambiguous—in every way far too limited. A massive mismatch looms between the information coming into our senses and the outputs of cognition.

In 397, Augustine of Hippo asked the same question, using the classical metaphor of the memory palace.

Great is this power of memory, exceedingly great, O my God, a spreading limitless room within me. Who can reach its uttermost depths? Yet it is a faculty of my soul and belongs to my nature. In fact I cannot totally grasp all that I am. Thus the mind is not large enough to contain itself: but where can that part of it be which it does not contain? Is it outside itself and not within? How can it not contain itself? How can there be any of itself that is not in itself?

What is the missing element, then? The mismatch between the information coming into our senses and the outputs of cognition arises from the brain reflexively filling in fleeting perceptual gaps with material that our minds "know" from memory belongs in the picture. Our perception always tends toward prediction: It anticipates what it is seeing. Much of our knowledge is instinctual and comes preprogrammed with our genetic code. A large part is appropriated from the experience of others through culture. And the rest is our personal experience. What we learn in the dozen or more years we spend in school is all content provided to us from our collective memory. Reading, writing, arithmetic, history, music, drawing—these are gifts of the generations, "derived from people," as Czeslaw Milosz writes, "but also from radiance, heights." The "spreading limitless room" of memory within each of us is limitless because we have access to the memory of humanity.

If the great feat of memory is to construct a model of the world that approximates reality closely enough—however we do it—the genius of imagination lies in using that model to create alternative orders and models of reality. Memory records the world *as so*. Imagination transposes it into the key of *as if*, transforming experience into speculation. That is why to lose one's memory means losing the future. Because imagination is memory in the future tense.

MAKE-BELIEVE

How does memory become imagination, and why? Imagination cannot come exclusively from memory and experience, as our scientists point out, because if it did, how do we account for the prodigal make-believe of children? Children have no experience to fill their imaginings with. During childhood the brain explodes with neuronal growth. The little neurons need to stretch and wriggle and find their place in the sun by rehearsing their future uses. These neurons spontaneously generate imaginary worlds as the mind builds its capacity to see patterns, make order and sense of a jumble of perceptions. Through observation and imitation, children begin to match their experiences of the world to what exists in their imaginary realms. They play dress-up, stepping into the imaginary clothes of imaginary adults. They act out what the clothes prompt them to do and work from a script they seem to know by heart without prompting. How this happens is one of the great puzzles of early development.

What we see in the imagination of children is a glimpse into the process of the nervous system learning the environment as the child grows. The nervous system rapidly builds extra mental and physical capacity just in case, and that capacity atrophies when not used. Use and experience begin to shape the consciousness of the individual. The neural connections that do not get used are radically pruned in early childhood. Some cognitive disorders such as autism may be caused in part by failures in the pruning process, so that some individuals maintain a physical sensitivity or model of thinking that is inappropriate for the world they grow up in. Windows of learning vary a great deal in their onset and duration. Children can learn many different languages fluently before puberty, but after puberty the capacity to learn a language diminishes. Learning a new language takes greater effort with age, and certain types of mastery may no longer be achievable. The same is true for many physical and artistic skills, such as gymnastics or playing the violin. We can learn these things at any age, but our minds and bodies are not as plastic as they were in youth.

CONJECTURE

Imagination in adults is quite different from what we find in children. It is more akin to conjectural thinking, the ability to predict based on incomplete information. The capacity for conjecture has been observed in children as young as six months old. They can notice correlations and even infer causal patterns. But a child's model of the world is full of enchantment and driven by desires. They tend to make predictions that invoke supernatural or magical forces and imagine consequences that are physically impossible. They simply do not know enough yet to make reasonable inferences about cause and effect, even though they have strong instincts about causal patterns. Conjecture requires a well-developed understanding of how the world works, whether you are trying to determine when to start cooking the eighteen-pound Thanksgiving turkey if the guests are to sit down to dinner at six, you set the oven at 450 degrees, lower it to 325 after thirty minutes, and when done, let the bird rest for thirty minutes before carving; how much thrust will send a satellite into high altitude orbit around the Earth; or how changes in mortgage rates will affect the real estate market and the rate of inflation.

Popular ideas about imagination are not deeply informed by biology, which is rather surprising given how routinely biological models are applied to explain other aspects of human nature. Instead, imagination is commonly and vaguely understood as a special kind of intelligence we use to "think outside the box" or "color outside the lines." This is why there is a brisk business in books, seminars, and technical manuals purporting to impart rules about how to think without rules.

This is a peculiar approach. Imagination is, as far as we know, unique to humans. But among humans, it is ubiquitous; we all have it. Imagination does not come with an enviable bonus gene that some lucky individuals are born with. It is an intrinsic property of human memory, the foundation of our mental time travel, capacity for problem solving, and conjectural thinking. Scientists hypothesize that certain animals can solve problems in ways that are similar in kind, if not degree, to humans. Chimps can scoop ants out of a hole with a stick and crows can manipulate a wire into a hook with which they coax some tasty treat from a notch in a tree. Orioles shred plastic bags and use the strips to line their

nests. These tasks require the transposition of something the
animal knows from one realm into a different scenario to produce
the desired effect. But the tasks observed by scientists do not
require deep temporal depth perception or mental time travel.
Only humans build bridges across deep bodies of water and send
their fellow men into the void of space to collect rocks from the
surface of the moon, because only we can imagine a series of
actions, each with specific consequences, each consequence
determining the path of the next step, and the series extending
over long periods of time. Such long-term planning is impossible
without collaboration among many people who engage in mental
time travel together.

Conjecture requires the ability to "combine images and experiences to construct an infinite number of future situations." When
conjectural thinking is used speculatively about what might
happen, we call it prediction or forecasting. In fact, it is the same
process as forensic detection, only in reverse. In the context of
science and engineering, what makes a prediction plausible is its
conformity to natural law. As the scientist Richard Feynman said,
"Science is imagination in a straitjacket."

The brain is a powerful simulator that runs a constant program
of "what if?" scenarios based on its memories and associations. As
insomniacs know, the mind's favorite diet consists of counterfactu-
als—what might have happened but did not. We can anticipate a
possible deviation from routine by rerunning scenarios of the past
and altering one factor or another to get different outcomes. People
like S. cannot engage in conjectural thinking or play with coun-
terfactuals because they do not recognize a routine or pattern as
such. Their disrupted memories cannot create patterns of events or
causal models. For S., running scenarios from the past could only
generate large amounts of recollected events that then become
confused with reality. People who suffer from serious trauma may
also have trouble with conjecture, though for different reasons.
In their cases, they are stuck in the past. No matter how many
times they run through a scenario that is somehow connected to
the trauma, even remotely, they always end in exactly the same

place they began. The trauma has robbed their memory of its innate plasticity. Many therapeutic techniques revisit traumas to open up the memory and modify the feelings it triggers—in essence, to rewrite the meaning of the remembered event by changing the emotions it arouses.

DISINTEGRATION: THE STORY OF IRIS MURDOCH

When describing how memory functions to lay audiences, brain scientists will often say that people *are* their memories. Our memories are not just an accumulation of data points about the past. They are the very fabric of the self, woven of our experiences, endowing us with time, place, personality, and identity in the world. The healthy mind is a master weaver, always at the loom fabricating and mending, trimming and reinforcing memories to make strong patterns of association readily available in a pinch. But in Alzheimer's, the weaver has gone mad and the moths have taken over. The fabric of memory disintegrates into smaller and smaller shapes and sizes, facts of a life story fall away with them, and at some critical point, so does the sense of cause and effect, of the self's continuity over time. No patterns are discernible in the tatters and rags. Meaning has disappeared. Events exist outside the context of time and place. The future is unimaginable and the present utterly without purpose. Whatever enters the consciousness of an amnesiac appears as a scrap of the past suddenly unmoored from its natural harbor in a life's narrative. It can feel frightening or vaguely sinister, floating before the amnesiac unattached, out of context, mocking with its familiarity but useless, empty of meaning.

Iris Murdoch (1919–1999) was a prolific writer of fiction and philosophy until she succumbed to Alzheimer's. As she lost her memory, her imagination atrophied and eventually she lost the ability to create the alternative worlds of fiction for which she was celebrated. The range of her interests, the scope of her imagination, her ability to forge connections between the abstract and the concrete had been Murdoch's singular gifts as a writer and thinker. But disease defies human logic. It was her very acuity of mind— her Iris-ness—that was torn apart by the brain shredder of Alzheimer's. In *Elegy for Iris*, the memoir by her husband, John Bayley, we see her disappear bit by bit. Bayley was able to redeem some of the suffering she experienced by writing so lucidly of the Iris he knew in the various states in which he knew her. He provides witness to the unraveling of the strong fabric of her powerful personality and mind, cherished by so many for its consistency and self-awareness.

Like S., Murdoch lacked temporal depth perception, though for diametrically opposed reasons. She was trapped in the hell of the eternal present with no exit into the future. For better or for worse, such memory damage in humans does not reduce us to dumb beasts who feel no anxiety. Describing Murdoch's descent into the dank fog of dementia, Bayley reports that "most days are for her a sort of despair, although despair suggests a conscious and positive state, and this is a vacancy which frightens her by its lack of dimensions." Her face became a "daily pucker of blank anxiety." Bayley recounts that for Iris, "Time constitutes an anxiety because its conventional shape and progression have gone, leaving only a perpetual query. There are some days when the question 'When are we leaving?' never stops, though it is repeated without agitation. Indeed, there can seem something quite peaceful about it, as if it hardly matters when we go, or where, and to stay at home might be preferable in any case."

No matter how old we are, our sense of well-being arises directly from our ability to imagine moving forward, into the future, with purpose, meaning, and some measure of choice over our fate. Because the past is the raw matter of imagination, people with Alzheimer's exhibit acute degradation of the imagination: no past, no future. Studies have shown that "in humans, the ability to imagine future events and consciously recollect one's past (episodic memory) is impaired in patients with damage to the region of the brain known as the hippocampus. Furthermore, new imaging studies show that some of the same brain areas are active during both planning and remembering in normal adults." The organ responsible for key aspects of memory is also responsible for conjecture and, by extension, key aspects of the imagination. But as S. shows, a richly embodied presence of the past does not constitute healthy memory per se.

We are like shadow puppets, the light of history illuminating our way by casting shadows of the past on our path, creating the necessary sense of familiarity we need to move forward into the unknowable future. Murdoch had no light from the past behind her to illuminate her way forward. She sometimes spoke of herself as "sailing into the darkness." What could she have meant, other than that the light of her memories, by which we navigate our lives, had been extinguished?

CULTURAL AMNESIA

The loss of collective memory is as devastating to cultural identity as the loss of personal memory was to Murdoch. In the wars of the last century, both civil and international, the destruction of cultural memory became a central strategy in subduing civilian populations. It began in the First World War and was taken up and elaborated by the totalitarian regimes that sprang up in its wake. The Bolsheviks emptied libraries, archives, museums, and churches of holdings that might serve to undermine their political goals and stand as mute witness of past times and past beliefs. These expropriations set the stage for the massive theft and destruction of Jewish and Slavic cultures by the Nazis, the Cultural Revolution of Mao's China, the destruction of cities during the Khmer Rouge's aggressive policies to effect agrarian socialism, and the destruction of the Buddhas of Bamiyan by the Taliban in 2001. During the War of 1812, when the Americans burned the parliamentary library in the capital of Canada, it was a big fat Yankee thumb jammed in the eye of the British Empire. It was not an assault on the cultural identity of noncombatants. The destruction of the Library of Congress the next year was simply retaliation. But the bombing of the National and University Library of Bosnia and Herzegovina in Sarajevo in August 1992 by the Serb artillery was not the destruction of a strategic military target. The library housed irreplaceable manuscripts documenting the Bosnian past. Destroying them was a deliberate attempt to wipe out the Bosnians' right to exist in Bosnia by wiping out the evidence that they had ever existed.

There is a deeper meaning for collective memory in the lives of S. and of his doctor, Luria. S.'s tenacious and pointillist memory was a painful yet poetic existential protest against the historical amnesia enforced on Soviet citizens. Luria must have known this when he wrote his book in 1965. S. (1886–1958) and Luria (1902–1977) lived through decades of dislocation and profound historical rupture during the thirty years they worked together, a fact never once acknowledged in Luria's account, published in the Soviet Union.

This was a curious time in the history of collective memory. The Soviet state apparatus operated under the radically utopian view that the foreordained outcome of human history is to be the triumph of communism and the withering away of the state. In this view, humans do not create their own futures; they are supposed to fulfill a preordained destiny. The ruling Communist Party was creating its own official past as party propagandists set about staging their own version of the inevitable future. The past had to be so arranged that it looked like there was only one possible outcome. They neglected the fact that history does not proceed in straight lines but in pushmi-pullyu fashion. It likes to go off in several different directions at the same time. Party historians were obliged, then, to launch a full-force assault on the past, erasing inconvenient facts from history books, wiping out people who gave credence to those facts, and fabricating alternative pasts that fit better with the story of history moving inexorably toward communism and the dictatorship of the proletariat. That is why Soviet citizens joked that in their land the future is certain; it is the past that is always changing. This bizarre temporal dislocation was widespread in Central and Eastern Europe during the decades of Communist rule. As a result, many people could feel cut off from real life, just as S. did. The Czech novelist Milan Kundera had the feeling living under communism that "life is elsewhere."

WITH AGE WE ARE ABLE to rely less on imagination and more on our experience. We would be fools not to. Yet without our imaginations, we quickly end up condemned to living in a world of old men, people who cannot be surprised, cannot accommodate new information, cannot feel sympathy for those who have not learned the same cruel lessons from life, and cannot summon the naïve optimism necessary to survive in a constantly changing world. Given the increasingly long life-spans of people in developed countries, this problem of the mind closing as we age presents serious challenges. How will society respond flexibly, inventively, optimistically to the increasing pace of change if we lose our imagination?

If for no other reason than this—to keep pace with ourselves as we rush in pursuit of the next new thing—the cultivation of our imaginations becomes more important with each successive stage of our lives. It also grows more important as the pace of technological innovation accelerates and we are called upon to update not just our software and hardware, but our habits and routines. Reducing unstructured playtime and cutting back on music and arts in elementary education is the wrong preparation for children who grow up in such a dynamic environment. Privatizing culture and entertainment and making it inaccessible to all but the well-off will rob adults of precisely what they need most to remain adaptable and emotionally accessible in a technologically advanced society. More than that, as globalization shrinks the world and makes encounters with people from different cultures more frequent, the cultivation of our moral imagination—the ability to see ourselves in another's shoes—becomes more important, not less. In the present age, we are told that knowledge is cumulative, expert knowledge abounds, and what counts is to add something useful to the sum of useful knowledge. It is hard to argue for the cultivation of the aesthetic imagination by saying that it is useful because it makes us better friends, mates, parents, citizens, workers. But it does precisely these things because it enhances our emotional intelligence. The aesthetic and moral imagination are intrinsic to the human condition. Without them we are less than human.

If, like S., we fail to form strong memories because we are constantly distracted, we will not find meaning or develop a sense of narrative in our lives. And if we fail to provide stewardship of our collective memory, we will be like Iris Murdoch late in life, unable to understand where we are in the present, without a past, and bereft of the future. The value of imagination goes far beyond foresight, conjecture, or prediction. It slows us down and gives us what Nabokov calls "consciousness without time." Now is the time to imagine how we will reconstruct our memory systems to accommodate abundance.

CHAPTER NINE

MASTERING MEMORY IN THE DIGITAL AGE

*I consider that a man's brain originally is like a little empty attic, and
you have to stock it with such furniture as you choose. A fool takes in all
the lumber of every sort that he comes across, so that the knowledge
which might be useful to him gets crowded out, or at best jumbled up
with a lot of other things, so that he has a difficulty laying his hands
upon it . . . It is a mistake to think that that little room has elastic walls
and can distend to any extent. Depend upon it there comes a time when
for every addition of knowledge you forget something that you knew
before. It is of the highest importance, therefore, not to have useless facts
elbowing out the useful ones.*

—SHERLOCK HOLMES, *A STUDY IN SCARLET*

MEMORY AT PRESENT

SOCRATES WORRIED THAT writing produces forgetfulness, for it is
an "elixir not of memory, but of reminding." He feared people's
"trust in writing, produced by characters which are not part of themselves" would erode their character and give people the mere "appearance of wisdom, not true wisdom, for they will read many things
without instruction and will therefore seem to know many things,
when they are for the most part ignorant and hard to get along with."

On the other hand, without writing (and photography and sound recording), we would be like Adam and Eve in Paradise: every morning waking up with fading memories of yesterday and no way to plan for tomorrow, let alone the day after tomorrow. Contrary to what Socrates held, we have done well enough in cultures rich with reminding devices, relying on durable media and multigenerational institutions to remember for us. We are Nature's generalists, ready to fit in wherever we can, adaptable because "our brains are the ultimate general-purpose organs, not adapted 'for' anything at all" except for learning to make do in whatever environment we find ourselves in. From cuneiform to computer chip, each innovation in knowledge storage has increased our fitness as a species, helped us adapt to new or hostile climates, find ways to feed our growing populations, lengthen our lives by curing or preventing diseases, and free up space in our brain attics for new questions to ponder.

A recent study on the "Google Effects on Memory" shows that we are adapting quite nimbly to having so much information at our fingertips. We are able to distinguish between the information that is available online, and what is not and that we therefore need to remember ourselves. "Just as we learn . . . who knows what in our families and offices, we are learning what the computer 'knows' and when we should attend to where we have stored information in our computer-based memories." Like graduate students who learn how to consume a dozen or more books a week in preparation for a life of intensive learning, we do not remember the content of stored information so much as where information is stored and how to find it. The authors shrewdly point out that "we have become dependent on [computers] to the same degree we are dependent on all the knowledge we gain from our friends and co-workers—and lose if they are out of touch. The experience of losing our Internet connection becomes more and more like losing a friend. We must remain plugged in to know what Google knows." Outsourcing more and more knowledge to computers will be no better or worse for us personally and collectively than putting ink on paper. What is important in the digital age, as it has been for the print era, is that we maintain an equilibrium between managing our personal memory and assuming responsibility for collective memory. In the twenty-first century that means building libraries and archives online that are free and open to the public to complement those that are on the ground.

How do we master memory in the digital age of abundance? It will start with retooling literacy for the digital age and updating public policies to ensure investment in long-term institutions capable of securing memory into the future. It will not happen in one generation, or even two. But it is time to lay the foundations and imagine the memory systems that will be in place—and by which we will be remembered and judged—when we are no more.

LITERACY IN THE DIGITAL AGE

In April 2010, the Library of Congress announced that Twitter would provide "public tweets from the company's inception through the date of the agreement, an archive of tweets from 2006 through April, 2010. Additionally, the Library and Twitter agreed that Twitter would provide all public tweets on an ongoing basis under the same terms" for the purpose of archiving them as historical sources. As such, Twitter is valuable less as a source of individual biographical information than in the aggregate as a database for large-scale analysis. One of the issues raised when the Twitter archive was donated to the Library of Congress was the matter of who owned the data and had the right to decide what to do with it. Many people were unpleasantly surprised to realize that they did not control their own Twitter streams. Some people also questioned the value of Twitter and its place in the national library, alongside the papers of George Washington and George Gershwin. A year later, though, headlines were ablaze with stories of the Arab Spring. Reports highlighted the role that social media platforms including Twitter played in helping to galvanize the attention of the world and organize protesters on the ground. It became increasingly clear that as a source for data mining at scale, social media data are valuable sources of information about social behavior, political actions, public health trends, and much else. We rely today on large-scale data analysis for most activities, from weather forecasting and figuring the odds on the Super Bowl, to making tax assessments based on the census and maintaining our air traffic control systems.

Large-scale data analysis demands a lot of data, and a lot of the data that come from us are used without our permission or even knowledge. Search engines know where a search originates, keep records of the searches, use those data for purposes never revealed to the searcher, and hand them over to authorities when required. We may find it extremely convenient when we make purchases on the web or search for information to have our machine "remember" where we search and what our shopping preferences are. We are less enchanted to realize those data can be accessed by intelligence agencies to trace our political affiliations and network of friends, and exploited by rogue hackers for identity theft or fraud. These practices are of concern to both citizens and consumers, but they also threaten to freeze advancement of knowledge that may benefit us all. The oceanographer James McCarthy warned in an address as president of the American Association for the Advancement of Science:

> More important than technical approaches [to data collection and mining] will be public discussion about how to rewrite the rules of data collection, ownership, and privacy to deal with a sea change in how much of our lives can be observed, and by whom. Until these issues are resolved, they are likely to be the limiting factor in realizing the potential of these new data to advance our scientific understanding of society and human behavior, and to improve our daily lives.

Literacy in the digital age means achieving autonomy over our choices of what we read and publish, who uses our data and how, and what happens to them over our lifetime and beyond. Literacy online, as in print, begins by learning to read with appropriate skepticism, being able to assess whether something we see is trustworthy or not, and being responsible in our use of data, our own and other people's. It means learning to interpret search results and how to select among them, knowing which links are "sponsored" by a company that pays for placement at the top and which not, understanding the basics of how computers encode information and display it, how data are created, collected, and used, and how to ensure the privacy and appropriate use of our data. We should be able to identify the source of information we read and evaluate its truth value, authority, and authenticity. We should be alert to a document's inconsistent use of tense, spotty noun/verb agreement, inexplicable changes in font, and redundancies as the signature of a text that has been cut and pasted from other sources. These are fundamental skills of reading and writing on the screen as on the page, to be introduced in early education and kept up to date over the decades.

Above all, literacy is about making informed choices about how to spend our time, the only asset in our abundant information economy that is truly scarce. We no longer have to seek information; it seeks us. It follows us wherever we go and, like a pack of yapping dogs, it begs for our attention. We need to reset the filters that control the flow of information ourselves. A well-kept digital brain attic has plenty of room for short-term tasks, but the pathways to long-term memories are always kept clear. Digital autonomy looks very different from the print model. It is not just about lowering barriers to access or outsmarting censors and copyright owners. It is about filtering data for value, creating for ourselves a time for deep absorption, and setting our machines to be on call when we need them but not to intrude on our privacy.

The quest for digital filters began almost immediately after the Internet and World Wide Web entered the public sphere. In the last century, everyone who went online experienced some degree of digital vertigo. The first filters to appear were search engines that promised to sort and organize information for us according to its potential value in answering the specific question we asked. After a short but intense period of competition between several commercial search engines (AltaVista, Lycos, Yahoo!, Ask Jeeves, among others), one company, Google, emerged as the predominant force, offering "to organize the world's information and make it universally accessible and useful." Soon after, social media sites like

Facebook, LinkedIn, and others (many, like Myspace, now roadkill on the information superhighway) began offering to manage all our social and/or professional activities. Internet commerce sites from Amazon to Zappos compete to monopolize our commercial transactions by offering everything to everybody everywhere ASAP. These commercial sites are designed to drive traffic to certain products and do so by bombarding people with distracting advertisements to get their attention and promising the irresistible reward of instant gratification. By extracting invaluable information from our use data, they create algorithms that predict our desires and streamline production facilities that offer to fulfill them even before we can articulate them—the "if you like this, you will like that" magic of user-driven algorithms. On the one hand, these shortcuts to gratification work for us because they save us so much time. On the other hand, we end up not with more freedom of choice but less, and the results can be easily gamed without our knowledge. The trade-off between choice and convenience is always there. A digitally literate person will be able to recognize when the trade-off arises and make a decision between choice and convenience for themselves. They will know, for example, what a cookie is and choose to allow them or turn them off.

In countries lacking free markets, Internet filters are imposed by political regimes. They govern which versions of the past and present realities are available to citizens. Where political powers try to control their population by controlling what they know, think, and believe, there is a lively business in censorship. Controlling the means of distributing information, be it censoring books or blocking IP addresses, is a critical first step. But really serious regimes need much more than censorship to be effective. They need to invent false pasts as well to calibrate expectations of the future. Circulating false information is a well-honed strategy of persuasion, used by political campaigns to smear opponents and by regimes against their enemies. In the Cold War, this came to be known as "disinformation." Such tactics are deployed in the digital age to wage ideological combat both at home and abroad, just as they were five hundred years ago. Among the earliest uses of printing presses was the production of both antipapal and anti-Lutheran propaganda. Ideological combat provided very good business for struggling early-stage start-up publishing houses.

In market economies, commercialization of "free" communication channels such as Facebook and Twitter sparks debate about a host of economic, political, and social challenges. Overlooked, however, are the potentially serious long-term implications for memory, both individual and collective. The long-term future of collective memory is not the business of commercial companies. By necessity they have a short time horizon and we cannot expect them to invest adequately in preserving their information assets for the benefit of future generations when these assets no longer produce enough income to pay for their own care and feeding in data archives. The problem for collective memory is not commerce's narrow focus on quarterly returns, deleterious as that is for any long-term planning. It is that commercial companies come and go. When they are gone, so, too are all their information assets. Unlike institutions established to serve the public trust, commercial companies have no responsibilities to future generations. The simple solution for preserving commercially owned digital content is for companies to arrange for handoffs of their significant knowledge assets to public institutions. The donation by Twitter of its archive to the Library of Congress is a signal example of the partnership between private and public institutions that will need to be the norm in the digital age.

AVOIDING COLLECTIVE AMNESIA

The abilities to reflect on one's own behavior, to transcend instinctual reactions, and to make thoughtful, well-reasoned choices are all enabled by a mind with deep temporal perception. Psychologist Daniel Kahneman has dubbed the ability to be aware of and reflect on one's own behavior, to move beyond purely instinctual reactions and make conscious choices, as "slow thinking," as opposed to instinctual reactions, "fast thinking." This distinction between instinctual and unconscious actions and those taken with deliberation is fundamental not only to decision making—the context in which Kahneman discusses it—but also in terms of creating healthy, long-term memories. Organizations are very good at slowing down human thought and reaction time to open up space for deliberation. "Organizations are better than individuals when it comes to avoiding errors, because they naturally think more slowly and have power to impose orderly procedures . . . An organization is a factory that manufactures judgments and decisions." Libraries, archives, and museums are necessary for the stewardship of our long-term memory because they are conservative—their job is to conserve—proceed in an orderly fashion to make judgments, and hold themselves accountable to their publics.

Who has the right to preserve digital content on behalf of the public—present and future? Do we own our personal data— biomedical, demographic, political—and can we control its use? If certain categories of data are private, then is metadata—data about data—about that data also private? These are not theoretical questions. Today, most of our personal digital memory is not under our control. Whether it is personal data on a commercially owned social media site, e-mails that we send through a commercial service provider, our shopping behaviors, our music libraries, our photo streams, even the documents on our hard drives written in Word or Pages—they will be inaccessible to us, unreadable in only a few years. A web-based wedding site will barely outlast the honeymoon and certainly will not be around to share with children and grandchildren in fifty years unless provisions to archive it are made by the wedding party now. A digital condolence book will disappear from the Internet long before the memories of those who still mourn the loss of that person dim. The documents on our hard drives will be indecipherable in a decade. We view our Facebook pages and LinkedIn profiles as intimate parts of ourselves and our identities, but they are also corporate assets. The fundamental purpose of recording our memories—to ensure they live on beyond our brief decades on Earth—will be lost in the ephemeral digital landscape if we do not become our own data managers. The skills to control our personal information over the course of our lives are essential to digital literacy and citizenship.

The marketplace of ideas is now conducted chiefly online. If something cannot be found online, chances are it will disappear from the public mind. The surest way to keep the records of the past readily accessible to the public is to migrate them to digital form, whether it is converting movies shot in Technicolor film to digital formats or digitizing eighteenth-century genealogical records and posting them online. These analog sources need to be preserved in their original physical formats. But digitization widens access to them at the same time it makes them accessible to many new publics.

Libraries, archives, and museums are making many of their collections available to the public online, but this vital service is hampered by chronic underfunding. The value we get from scanning is far greater than digital access alone. We have discovered that a book, manuscript, map, or painting with one value in arti-factual form can have multiple novel values when converted into data. One example of digitally augmented value is the corpus of seemingly mundane ships' logs, written in various crabbed hands over the course of centuries, each documenting the details of overseas voyages. Of course, they have long been important for people writing the history of exploration, trade, and all matters maritime. Recent conversion of these logbooks gives them new value as a database and makes invaluable historic evidence about the climate and marine ecologies accessible to computer analysis. Who could have guessed the value of the logs' detailed information about weather, ocean currents, schools of now-rare fish that once were abundant? Historical information about the climate is difficult to come by and worth its virtual weight in gold for those trying to understand long-term patterns of stability and fluctuation in oceanic and atmospheric conditions. That information has been lying dormant in old documents and logbooks for centuries, accessible only to people who visit the archives to study them. Until now, it has been impossible to read them at scale, to understand what the entire corpus can tell us about long-term trends.

The forensic shift of the nineteenth century led to assembling large-scale collections of unimaginably diverse objects that hold valuable information in often fragile and unwieldy objects. Medical history museums have extensive collections of human and animal tissue samples that document the history of disease. Natural history museums have drawers full of birds, bugs, and bones that now can be sampled for DNA to yield information about genetic relationships, often rewriting the genealogical trees of life. The Avian Phylogenomics Project is sequencing the genomes of forty-eight birds from forty-five avian species to construct a family tree for birds. Over 60 percent of the tissues to be sampled will come from natural history museum collections. Institutions once derided as antiquated warehouses full of stuffed specimens turn out to be the Fort Knox of the genetic research era. Some astronomical observatories hold collections of glass plate negatives of the night skies taken in the nineteenth and early twentieth centuries. Here we find records of unique celestial events, such as supernovas and asteroids. One series of images led to a discovery that could never have been anticipated at the time the plates were exposed, the hitherto unknown source of energy we now know as a quasar, first verified in 1962 by studying seventy-year-old glass plate negatives.

Digital information processing not only allows new uses of old sources. It also rescues fragments of memory once thought irretrievably lost. Technologies created at great expense to advance one field of knowledge can yield unintended benefits for others. Take the case of the Large Hadron Collider, an atom smasher that in 2012 detected the Higgs boson, a subatomic particle predicted to exist but never observed. The imaging technology that recorded traces of the elusive Higgs was adapted to access the authentic voice of Alexander Graham Bell and others trapped on unplayable recordings by "visualizing sound," capturing transient acoustic waves on materials that "record" them.

Sound recording is a recent technology, the first recording made in 1860. Despite its youth, in many ways audio is far more vulnerable to decay and loss than parchment manuscripts that have survived for two thousand years. The materials required to fix acoustic vibrations onto physical substrates and the elaborately engineered playback systems required to create sound waves from these substrates are uniquely fragile. Sound waves can be represented three-dimensionally by grooves incised on a flat disc, wave patterns impressed on clay cylinders coated in wax, or rolls wrapped in tin foil stamped with wave. The problem is that to hear the sounds, a stylus needs to travel across the grooves, and each time it does so, it wears down the disc of lacquer, plastic, aluminum, shellac, or cylinder covered in wax or foil. Some are so worn they cannot be played. Others have little deterioration of their grooves but are broken into pieces. What we need to hear sounds engraved on objects too fragile to subject to physical replay is to capture the information on the surface without making physical contact with it.

In 2000, an experimental physicist at the Lawrence Berkeley National Laboratory accidentally learned about the at-risk audio problem. Carl Haber was at work "developing devices to image the tracks of subatomic particles which would be created at the Large Hadron Collider." What about "playing" the disc by creating a map of the surface of the disc and converting the image into sound waves? No contact, no damage. The sound could even be cleaned up, amplified, and noise reduced. Working with sound recording engineers and preservationists, Haber and his research partners developed a cluster of technologies, called IRENE (Image, Reconstruct, Erase Noise, Etc.), that image sound in both 2-D and 3-D (the latter useful for cylinders). It is now possible to hear the voices of Native Americans performing tribal rites that had been lost to their descendants and the voice of Alexander Graham Bell himself as he tests out his new machine. The forensic imagination means that there are now almost limitless possibilities for the extraction of information from any piece of matter, no matter how fragile.

BIG SCALE AND CONTESTED RIGHTS

The old paradigm of memory was to transfer the contents of our minds onto a stable, long-lasting object and then preserve the object. If we could preserve the object, we could preserve our knowledge. This does not work anymore. We cannot simply transfer the content of our minds to a machine that encodes it all into binary script, copy the script onto a tape or disk or thumb drive (let alone a floppy disk), stick that on the shelf, and expect that fifty years from now, we can open that file and behold the contents of our minds intact. Chances are that file will not be readable in five years, and certainly far less if we do not check periodically to see that it has not been corrupted or that the data need to be migrated to fresher software. The new paradigm of memory is more like growing a garden: Everything that we entrust to digital code needs regular tending, refreshing, and periodic migration to make sure that it is still alive, whether we intend to use it in a year, one hundred years, or maybe never. We simply cannot be sure now what will have value in the future. We need to keep as much as we can as cheaply as possible.

But what is possible? Such a model defies the fundamental proposition that whatever we do in the digital realm needs to be able to scale up almost infinitely. To be able to store petabytes and exabytes and other orders of magnitude of data for long periods of time, we will have to invent ways to essentially freeze-dry data, to store data at some inexpensive low level of curation, and at some unknown time in the future be able to restore it—to add water and grow, not unlike the little party favors made of tightly wadded pieces of colored paper that when you put in a glass and fill with water expand to become villages or buildings or gardens or animals. Although this represents a serious challenge for computer science, it may be possible to store not the full file, but the instructions on how to re-create the file, just as the genome does not store an animal or plant itself, but the instructions on how to grow a partic-ular animal or plant. Until such a long-term strategy is worked out, preservation experts focus on keeping digital files readable by migrating data to new hardware and software systems periodically. Even though this looks like a short-term strategy, it has been working well for relatively simple text, image, and numeric formats like books, photographs, and tables for three decades and more.

The need for scale presents a second problem, of course, and that is how we make sense of all these data. This is a hard problem, but not insuperable. It will be solved by machine intelligence that identifies patterns in a set of data and thereby determines the context in which the data "make sense." (This is essentially how brains figure out what they perceive, by comparing real-time perceptions with those stored in memory to determine what something means and what its significance is.) Only machines can read computer data, and only lots of machines can read data at scale. But machines are extensions of us, tools that are simply the means to ends of our choosing. We are the ones who will design, build, program, and run the machines. We will decide how they get used, by whom, and for what purposes. And we will be the ones who must make sense of the results they give us. That said, we can imagine making sense of declarative memory—facts, figures, and whatever can be rendered in binary code—as immensely complicated but still tractable. How machines encode affective memory, denoting emotional salience, ambivalence, ambiguity, even something as simple in natural language as a double entendre, is a different order of complexity.

Beyond the problem of sheer scale, there are formidable social, political, and economic challenges to building systems that effectively manage an abundance of data, of machines, and of their human operators. These are not technical matters like storage and artificial intelligence that rest in the hands of computer scientists, engineers, and designers. They are social. Digital infrastructure is not simply hardware and software, the machines that transmit and store data and the code in which data are written. It comprises the entire regime of legal and economic conditions under which these machines run—such as the funding of digital archives as a public good, creating a robust and flexible digital copyright regime, crafting laws that protect privacy for digital data but also enable personal and national security, and an educational system that provides lifelong learning for digital citizens. We need to be competent at running our machines. But much more, we need to understand how to create, share, use, and ultimately preserve digital data responsibly, and how to protect ourselves and others from digital exploitation.

In similar fashion, each organization, society, laboratory, photography studio, medical practice, architectural firm, law practice, financial services company, and so on is now obliged to manage its own corporate files. In professions with legal and fiduciary obligations to maintain specified types of data, data management and archiving systems are standard operating procedure now. Commercial firms—particularly those in the business of selling "creative content" (music, films, video games) protected by copyright—have complete control over the long-term fate of what they produce, despite what is clearly in many cases the strong public interest in preserving that content and making it available after its commercial value is fully exhausted. At present, such companies have no financial incentives to hand off their cultural assets to an institution that can ensure their long-term public access. Economic and tax policies can be used in these cases to ensure the continued growth and support of public domain cultural materials.

The copyright law in the U.S. Constitution was created by the founding legislators to provide incentives for creators to circulate their ideas in the marketplace. In 1787, the authors of the copyright law made provisions "to promote the Progress of Science and useful Arts, by securing for limited Times to Authors and Inventors the exclusive Right to their respective Writings and Discoveries." They introduced financial incentives to publish by granting to the copyright holder exclusive rights to disseminate their works for fourteen years. Monetary incentives were deemed more democratic than the patronage of the church, the aristocracy, or royalty. Copyright law then coevolved with new technologies and new institutions, usually a step or two behind the facts of innovation. The roles of public access to information through libraries and archives also evolved and became more critical as the United States matured and absorbed more immigrants into the population. In light of the increasing importance of libraries to the economic, political, and cultural life of the nation, the copyright code was updated to grant exemptions allowing libraries to lend copyrighted content to the public and to create preservation copies to ensure continued access.

Yet in the digital age, the fundamental mission of libraries and archives to preserve and make knowledge accessible is at risk because there is no effective exemption from copyright law that covers the specific technical preservation needs of digital data. It is unrealistic to assume that in market capitalism, publishers, movie studios, recording companies, and other commercial enterprises will preserve their content after it loses its economic and commercial value or becomes part of the public domain. This is what libraries do. Nor is there a provision in the copyright code, comparable to the one that exists now for printed materials, that allows for libraries to lend their books, films, and audio recordings digitally.

The World Wide Web is not a library. It is a bulletin board. That was how it was originally designed by the computer scientist Tim Berners-Lee—to be a neutral medium of exchange—and that is what it continues to be. It will be a challenge to re-create the traditional public library online, because a public library exists in large part to provide access to contemporary copyrighted materials. The current copyright law, in particular the provisions that allow public libraries to lend their materials, does not apply to digital books or any other content in digital form. This means that any material that is under copyright, even if held by the library, cannot be put online without express permission of the copyright owner. This includes for all intents and purposes everything, in all formats, including audiovisual material, that has been published in the United States since 1923. Given the recent history of copyright extension, this means the twentieth century may be dark for quite a long time. Any material created since 1923 and still under copyright protection in 1998, when the Sonny Bono Copyright Term Extension Act was passed, will not enter the public domain until at least 2019 and possibly later than that.

The scope of the necessary modifications to the legal regime of copyright and contract law is clear enough, though they will be slow to effect because they require balancing the interests of private parties and the public domain, an action that the U.S. Congress must take to effect the changes. Far less clear are the economic models that will support the value of information—both short-term and long-term—including the cost of creating, using, and keeping it available for use. Much of the content on the open web, for example, has been contributed "for free," which really means that the user pays no transaction fee directly to the creator to use the content. It does not mean that the expert who contributes an article to Wikipedia, the scholar who writes a blog, and the member of the general public who joins the crowdsourcing project to curate old menus put online by the New York Public Library do not contribute valuable time and labor. The very norms that decide who owns what information—even which information is public and private—are still a matter of great debate. Some have suggested that each person who contributes information to the open web should be reimbursed by some micropayment every time it is used for commercial purposes. In truth, there is no consensus on how to determine the economic value of online information yet (other than the reliable rule that data are worth what someone will pay for them). This is particularly true of the kind of data contributed by users who upload information freely for one purpose—contributing to a review site such as Yelp or placing a classified ad on a bulletin board such as Craigslist—to have that information used by a third party for different purposes, such as demographic analysis that might be marketed. The privacy policy of these sites will tell users what they can expect to happen with the data they contribute, so the choice to add data or not is in the hands of the contributor in the end. That said, the scope for abuse is vast and may over time discourage the contribution of content for public dissemination.

IN THE MEANTIME . . .

Building resilient and ubiquitous digital memory systems will take time. It will require concerted investments of human and financial capital to model and test approaches. There will be near misses along the way, but failure can be very instructive. There will be social, political, economic, and legal wrangling as people and corporations scramble to secure rights and revenue streams before they even know which business models will support growth and which will crush it. We are still in the early days of the digital era.

The best, if not the only, way to understand the powers and limitations of the technology is to use it. In the meantime, until the arrangements work themselves out, the opportunity for individuals to make a difference will be almost unlimited.

Because of the distributed nature of the web, it is easier than ever to be a collector. There are no culturally agreed-upon norms regarding which content has value in the digital age and which does not, so individuals and small organizations that collect at scale today will help to determine the value and authenticity of today's content in the future. There are new populations of digital historians, librarians, and archivists who are aggressively collecting and preserving online information they think has long-term value. One of the earliest examples of such foresight began within days after September 11, 2001. People were rushing online to express their feelings and share information about events as they witnessed them. A pioneering group of historians at the Center for History and New Media at George Mason University immediately set up a site to solicit personal testimonies about the attacks of 9/11. This includes raw, rare, and invaluable eyewitness testimonies. This early example of crowdsourced documentation was gathered and put in good order by historians at the center, then transferred to the Library of Congress for posterity. This was the first digital collection acquired by the Library of Congress, most fittingly a collection born of allying private and public purposes and entrusted to the stewardship of a publicly supported library.

Great collections are made by great collectors. Between 1996 and 2014, the Internet Archive collected and preserved over 450 billion web pages. To a large extent the archive serves as a preview of what research libraries in the future may become, collecting and preserving for future access vast amounts of publicly available digital content. In addition to preserving significant portions of web communications, the Internet Archive enables people to archive personal digital collections. It also allows free uploading of digital files, as well as scanning books, films, television, and all manner of analog materials to broaden access to them. National libraries, archives, and research institutions around the globe that are in the collecting business have created a consortium to coordinate and broaden their scope of digital collecting. But so far no other organization has accomplished what this small nonprofit organization has. The Internet Archive is a classic start-up—nimble, opportunistic, driven by strong ambition and breathtaking vision, the very model of the Internet culture itself in its not-for-profit mode.

Crowdsourced collecting depends on the open web, and like the American West in the nineteenth century, what was once a frontier of open range is closing fast. Not only is the web being fenced in by commercial entities, but to an alarming degree, it is being ignored and altogether sidelined as more digital content is distributed through closed proprietary systems and apps that are wedded to specific operating systems, software, and hardware, a model pioneered by Apple for music on mobile devices and quickly imitated by Amazon for e-book readers. To some extent, these changes are part of a battle for market share among several major technology companies, and the market will eventually sort out which services consumers want the most, how they want them delivered, and what they are willing to pay for them. Already we have seen mobile devices take the place of desktop computers, laptops, cameras, music players, landline telephones, maps, atlases, watches, journals, and newspapers for delivering information.

In Jefferson's vision, access to organized knowledge is necessary to promote the progress and well-being of humanity. In the developed world, market capitalism plays important roles, but long-term investment in the public good is not one of them. Google boasts that they organize knowledge for the world. But the vision of Jefferson and the Founders proposes that the organization of access to knowledge is to be a public utility, wholly owned by the people for the purpose of self-government. Unless there is a handoff made between private entities that have the power to create, disseminate, and own content on the one hand and the long-lived nonprofit institutions capable of providing stewardship on the other, it will be hard to avoid collective amnesia in the digital age.

PART THREE

WHERE WE ARE GOING

At the present moment, something new, and on a scale never witnessed before, is being born: humanity as an elemental force conscious of transcending Nature, for it lives by memory of itself, that is, in History.
—CZESLAW MILOSZ, "ON HOPE," 1982

CHAPTER TEN

BY MEMORY OF OURSELVES

Time and accident are committing daily havoc on the [documents] depos-
ited in our public offices. The late war has done the work of centuries in
this business. The last cannot be recovered, but let us save what remains;
not by vaults and locks which fence them from the public eye and use in
consigning them to the waste of time, but by such a multiplication of
copies, as shall place them beyond the reach of accident.
—THOMAS JEFFERSON TO EBENEZER HAZARD, 1791

THE PREDICTABLE UNPREDICTABILITY OF THE FUTURE

MUCH AS THOMAS JEFFERSON had expected, today scientists, engineers, and technologists play leading roles in advancing knowledge and making it useful in our daily lives. In the twenty-first century, we command powers to release massive amounts of energy by splitting an atom and alter the script of life by splicing genes. Just as Jefferson anticipated, every advance in knowledge has brought with it new powers and new responsibilities. He built a library to ensure access to knowledge so that we may better govern ourselves. But we outgrew Jefferson's library, with its driving ambition to comprehend all of human knowledge by gath-ering printed volumes into one place, over a century ago. As the volume of information demanded by technology proliferates, it is increasingly difficult for us to know what we know, let alone take responsibility for it. How will digital memory shape our world in the next fifty years?

Digital technology is changing as we breathe. It will continue to change as scientists discover more about natural and artificial memory, engineers continue to design, test, and redesign hardware and software, and developers bring more products to market. At present, we are like adolescents in the throes of life's most awkward age, aware of the increasing powers we possess but with little idea how to use them, let alone control them. And like adolescents, we find the future and its possibilities so enticing that we seldom look back, neglecting or dismissing the past. The first generation of digital natives has not even reached middle age, the time when people can reflect, take stock of what they have done, what remains unfinished, and how they wish to be remembered when they are no longer among the living. Only decades of living with digital memory will reveal how reading on a screen differs from reading on a page, how digital audio recording affects our acoustical sensibilities, and how the inescapable ubiquity of information that chases us rather than the other way around alters our habits of thought.

The affect of overload will quickly fade away. For everyone born in the twenty-first century, the digital landscape is a given, the natural state. Being a digital native means the abundance of information does not feel like an overload. We grow up habituated to it and instinctually filter it. A quick glance around any sidewalk in any town shows the most common form of filtering today is the metaphorical earbud: tuning in to the information channel of your choice—be it texting, web surfing, listening to music—and tuning out everything else. The earbud restores autonomy to us in a noisy environment. Memories of the status quo ante—life before computers—are fading and will grow as dim as our collective memory of telegrams. The mixture of excitement and dread we felt when the doorbell rang and a man in a Western Union uniform handed us an urgent message from someone faraway is a mere footnote in history.

Information inflation is old news, a predictable and ultimately passing wave of turbulence in the wake of new information technology. The invention of movable type catalyzed changes in consciousness and behaviors that shocked, scared, and thrilled contemporaries. From the 1450s to the 1530s and beyond, the knowledge landscape was in a state of unremitting upheaval. Only in retrospect do we see it as a transition, as manuscripts continued to be copied by hand even as printed matter grew in volume. Yet by Montaigne's day, when two generations of print natives had come and gone, Europeans were troubled less by the quantity of the books than by their quality. The sheer volume of print was to them simply a fact of Nature. And five centuries on, the changes printed matter wrought, for good and for ill, are invisible to all but historians. The innovations of print have been completely naturalized. Most of us do not even see the printed book as "information technology." And yet in its time, the book was considered so potent a technology for disseminating ideas, so dangerous an agitator of emotions, that political and religious authorities tried to stop women and slaves from reading.

The benefits of digital memory are still unfolding before our eyes—faster communication among family and friends, empowerment of political activists to organize, economies of scale in manufacturing and distribution, increased access to cultural resources to educate, not to mention easy comparison shopping and price matching. The disadvantages of the technology have already been diagnosed and decried—frighteningly efficient surveillance, undetectable theft of private data, and of course, the loss of incalculable "hours of productivity" as we play games and fall down endless rabbit holes of online information. How do we decide what to keep and what to lose? On the one hand, our experience with digital data's here-today-gone-tomorrow quality advises us to follow Thomas Jefferson's advice to Ebenezer Hazard: Make lots of copies and spread them around as insurance against loss. On the other hand, experience also tells us that having too much can be as bad as having too little. In the wake of Edward Snowden's revelations of massive data sweeps by the National Security Agency

(NSA), we learned that the agency collects and holds such quantities of data—how much data it holds is classified—that they often cannot make sense of it for the purposes intended. A former NSA scientist said that staff find themselves "drowning in data" and often miss what might be significant. The sheer quantity of data makes it inherently unmanageable and impossible to detect, let alone stop, the abuse of data—our data.

What is at stake in the next fifty years is clear: a data universe that will provide unhampered access to information, protect our privacy, and remember us when we are no more on this Earth. How it will turn out calls for speculation, not prediction. Daniel Kahneman points out that we are prone to overconfidence in our predictions about the future. "Our tendency to construct and believe coherent narratives of the past makes it difficult for us to accept the limits of our forecasting ability. Everything makes sense in hindsight . . . And we cannot suppress the powerful intuition that what makes sense in hindsight today was predictable yester-day." The evolution of collective memory from the cuneiform's solution to accounting to the book's ability to record the inner-most secrets of the heart looks like a logical progression—not simple, yet somehow inevitable. It was not. That said, certain trends now shaping our future are clear.

Our twin aspirations—to be open yet to protect privacy, to embed democratic values in our digital code to support the public good while fostering competition and innovation in the private sector—will clash repeatedly. In addition, we face the risk of distraction—being enticed and seduced by the pull of so much novelty. We also face the risk of amnesia—of discounting and dismissing the past. If distraction prevails, we suffer the fate of S., who accumulated vast amounts of factual data and lost all sense of purpose and meaning. If amnesia prevails, then we casually let the full freight of human memory decay, bid farewell to six thousand years of humanity's written record, and, like Iris Murdoch, starve our imagination of its very sustenance.

But we are adaptable creatures. The default of digital memory—not fixed to durable objects, not constrained by the limits of time and place, never truly permanent—will shape our model of the world and vision of the future in three significant ways. First, as we learn more about the processes by which the past manufactures the future, we will use that knowledge to accelerate the growth of digital infrastructure, provide more secure stewardship of collective and personal memory, and reduce the risk of losing our past. Second, we will gain understanding of the complex processes by which Nature organizes itself and harness them to organize our data universe. Third and most important, we will outsource more tasks of memory, search, and retrieval to machines that outperform us and thereby make room in our brain attics for the cultivation of our emotional and imaginative powers—powers indispensable to thrive in a machine-driven world. None of this is predestined. But all of it is possible if we ensure long-term access to humanity's memory in a wired world largely open and equally accessible across the globe.

WHAT CAN WE AFFORD TO LOSE?

The good news about digital memory is that we do not need to lose very much information at all. On the contrary, we can generate and use unlimited amounts of data. True, we do not yet know how to preserve it for indefinite periods of time. Our capacity to search and filter the data universe is still primitive compared to what it will be in twenty-five years, let alone fifty. The ability to extract great amounts of information by imaging obsolete formats such as sound recordings on wax cylinders or to read the genetic data of old life forms will grow apace. The argument for keeping as much data as possible is almost unassailable, even though our ability to care for and manage data will always lag behind our ability to generate it.

The logic of biological memory is radically conservative. Successful species have enough redundancy in reproduction to withstand all but the worst of luck. Birds breed more in fat years and less in lean, but in all years produce more than they can feed in case a predator invades the nest, a chick eats a sibling, or a storm knocks the lot of them to the ground. They cannot survive a sudden change in habitat (losing a forest to logging or losing eggs by feeding on insects poisoned with DDT) any more than dinosaurs could survive the impact of an asteroid sixty-five million years ago. The code for each creature is always in peril if not copied copiously, redundantly. The moral is that in Nature, more is better than less.

And so it is with our artificial memory. The more fragile the medium, the more redundancy we need. Nothing we have invented so far is as fragile as digital data. We began our attempt to cheat death by creating mighty artifacts of clay, stone, paper, and parchment that outperformed our memory by hundreds and thousands of years. Now we create storage media that maximize volume, not durability. The Sumerian scribes looking down on us from their imaginary perch in space-time would be surprised at how far we have gotten in documenting the world and its many transactions over time, how far beyond accounting, epics, and prayers we have extended the memory of humanity, and how many people can read, write, and circulate their ideas across the globe instantaneously. They would marvel at the trade-offs we so lightly make between volume and durability.

But we may not have to make such trade-offs forever. We are entering now into an experiment with memory that was not even imaginable until a few decades ago—to take the first, most compact, and most enduring form of memory, the DNA molecule, and encode it with digital data. A research team in Switzerland has "devised a system that encapsulates and protects DNA strands in silica glass. The team also included redundancy codes to correct errors that arise when writing, storing, and reading that data." They tested storing the data in conditions equivalent to 10 degrees C for two thousand years and rendered them readable again. This was a successful proof of concept and significant step in developing artificial memory. That said, it will be some time before DNA is the standard for long-term preservation, if ever. Tampering with codes for self-replicating creatures raises a host of ethical issues that need wide public discussion before we consider digital preservation "solved" by DNA. But biological storage has been in the sights of scientists ever since it dawned on them that the limit to storage on silicon will be reached sooner rather than later.

And it cannot be soon enough. We are an incorrigibly curious species. Our appetite for more and more data is like a child's appetite for chocolate milk: Our eyes will always be bigger than our stomachs. In 2014, we embarked on two projects to map and model the human brain—one in Europe, the other in the United States. One scientist, when asked about the challenge facing large-scale brain-mapping initiatives undertaken by the European Commission's Human Brain Project and the U.S. BRAIN Initiative, said, "It makes Google's search problems look like child's play. There are approximately the same number of neurons as Internet pages, but whereas Internet pages only link to a couple of others in a linear way, each neuron links to thousands of others—and does so in a non-linear way." Given that one cubic millimeter of brain tissue will generate two thousand terabytes of electron-microscopy data, and the brain comprises on average eleven to twelve million cubic millimeters, where are we going to store all this information? So rather than less, we are certain to collect more. The more we create, paradoxically, the less we can afford to lose. Our entire tech-intensive economic, political, and cultural infrastructure is crucially dependent on secure and reliable data about everything from our online tax filings and bank deposits to the locations of toxic waste burial sites and nuclear bomb codes. Secure access to data in turn demands reliable sources of energy.

THE TIME FRAMES OF THE FUTURE

Free access to information is the sole guarantor of self-rule. Ignorance and secrecy are fundamental threats to freedom because they compromise our autonomy and freedom of choice. This is what Jefferson and his peers believed and why they established a national library funded by the public purse. Through the program of copyright registration and deposit, the Library of Congress preserves the record of American thought and creativity for present and future generations. In the Library of Congress, as in all public libraries, readers have open access to information while their privacy and the record of what they searched is protected. For most people, though, the most important library is not in the nation's capital, but in their local community. The growth of the Internet should not come at the expense of public libraries. Instead, we should be thinking about what services the Internet provides more effectively than a trip to the local library, and what services only local libraries can offer. Instead of thinking about the Internet and local libraries in the category of either/or, we should envision them in a symbiotic both/and relationship.

The web has the scope of a comprehensive library, but it lacks a library's rules of access and privacy. Much web content is inaccessible behind paywalls and passwords. Readers leave vivid trails of where they have been online in their browser history. To reinvent something like the Library of Congress or Alexandria online, we would begin with an Internet that provides easy access to information, make it searchable by something more comprehensive than Google, and add the crucial back end of a network of Internet Archives to ensure persistence of data. Readers and researchers would have use of cheap, ubiquitous fiber connection, regulated as a utility to ensure equitable access. The reading room of the public library, open to all and allowing privacy of use, would now be complemented by similar spaces on the Internet.

This may seem a complex vision to achieve. But the vast system of public and private libraries and archives that span the globe today and have worked so well in the book world is no less complex. The issues outlined here, from digital copyright, data privacy, and online library lending to mass digitization and digital preservation, have mobilized many information professionals to concerted action. For better or for worse, spectacular data breaches, government spying programs, contentious copyright suits that reach appellate courts, and the loss of personal and business data that happens every day are making the general public aware as well.

That awareness creates demand for secure, reliable digital infrastructure, the first step in addressing these challenges at scale.

The most visible component of the digital information infrastructure to most people is Internet search, which is essentially the domain of commercial technology giants. Search is difficult and expensive to improve. The good news is that archiving is among the least expensive pieces of the digital infrastructure. Google employs over 50,000 people (though how many work on search-related issues is unclear). The Internet Archive is a not-for-profit enterprise, supported through individual donations and foundation grants, that employs 140 people. By the end of 2014, the Internet Archive had archived "20 petabytes of data—including more than 2.6 million books, 450 billion web pages, 3 million hours of television (including 678,000 hours of U.S. TV news programming) and 100,000 software applications." It hosts over two million visitors every day and is one of the world's top 250 sites. Its search mechanism, the Wayback Machine, locates URLs, not words, and so is quite limited in speed and function compared to the search engines most people use to navigate the web. Even so, people find their way there in search of history—often their own—or following links from Wikipedia, the sixth most heavily trafficked site on the web. A study from 2013 reports that about 65 percent of people search the Wayback Machine for pages that have disappeared from the live web. (Today a web page lasts on average for one hundred days before changing or disappearing altogether.)

The Internet Archive, commodious as it is, faces constraints in collecting primarily for financial reasons. But if replicated across the planet and robustly funded, a global network of Internet archives could scale up to cover all languages and types of information. Without such a publicly accessible archive, the history of the twenty-first century will be riddled with large-scale blanks and silences, rendering our collective memory as unreliable as someone with Alzheimer's.

Providing access to archived data and ensuring that things are readable over long periods of time is a complicated technical problem, but largely tractable. But no data, no access. Other than the fact that preservation yields long-term rewards, and most technology funding goes to creating applications that yield short-term rewards, it is hard to see why there is so little investment, either public or private, in preserving data. The culprit is our myopic focus on short-term rewards, abetted by financial incentives that reward short-term thinking. Financial incentives are matters of public policy, not natural law, and can be changed to encourage more investment in digital infrastructure.

Curiously, both Google and the Internet Archive date back to the end of the last century, when the web was still small, neither interactive nor social media existed, and the young world of technologists was full of people with utopian ambitions to collect, organize, and serve the world's information. The search firm and the archive both thought big. But they were thinking across fundamentally different time horizons. For short-term access, Google cannot be beat for volume and performance. (Let Google serve here as an example of commercial search.) In fifty years Google itself will be unrecognizable. It will either have been replaced, or it will have morphed into a completely different company to stay competitive. Search itself may be regulated as a public utility, once enough people in Washington realize that information is the new energy. As long as search companies and content purveyors such as Apple and Amazon operate in a highly competitive space, their success will depend on staying closed, enveloped in the self-protective culture of NDAs (nondisclosure agreements), using customer data to sell ads, and licensing content for short-term uses.

The Internet Archive is in a very different business, one that is always at risk of failure, but can never be superseded. For archives to succeed at digital preservation, the web and the Internet must be open, not closed. Rather than competing against each other, archives can increase their chance of succeeding as they multiply and cooperate. An archive's assets are never deleted or expunged for political, economic, or even privacy reasons. Access to the data can be controlled for any number of compelling reasons, from protecting an individual's privacy to embargoing the location of endangered species or archaeological sites to protect them from predation. Paper-based archives routinely observe embargoes for limited times—say, until after the death of those mentioned in the archives.

Digital memory will determine how we know our own past, are exposed to the diversity of other human pasts, and how people remember us when we are gone. A long-lasting conversation among generations separated by hundreds and thousands of years was interrupted and almost totally silenced between the fall of the Roman Empire and the Renaissance. Since its revival and conversion to print, its influence has extended far beyond its origins in the Mediterranean basin. Now, with the Internet, we can continue that conversation as we digitize materials from the past, but also broaden it across multiple languages and civilizations with different historical experiences and expectations of the future. Though the dream of archiving Internet content was born of idealistic and utopian ambitions, in reality, it is evolving into one of the most efficient and economical ways to ensure the continuation of that conversation. We can leverage the proliferation of copies of things on the web to increase their chance of survival. True, with so much redundancy there will be a problem of version control. Solving search across so much information will also be nontrivial, as computer engineers are fond of saying. But these are technical matters, no matter how complicated.

The real challenge lies in mobilizing people to set aside their temporal chauvinism and focus on the long term of the past and the future. Despite the futurist fervor endemic in early digital decades, we now see the past is more valuable to us than ever. As Jefferson understood, it is because we care about the future that we have to know our past—all of it. That might be why the Long Now Foundation, located in the heart of the Bay Area's technology industry, was formed in 1996 to encourage long-term thinking. (Anticipating another Y2K challenge, the foundation always uses five digits for years, so it was, more accurately, founded in 01996.) The mission of Long Now is to encourage our species to think about our ambitions and challenges in increments not of decades, centuries, or even a few millennia, but ten thousand years. Their goal is to remind us that our actions have consequences. Their programs foster experiments in thinking through future scenarios and anticipating the responsibilities we must assume for the consequences of our actions over long periods of time. Among their projects is the documentation of endangered languages (the Rosetta Project), efforts to preserve and restore endangered or extinct genetic codes, including those of the passenger pigeon, the black-footed ferret, and the woolly mammoth (the Revive & Restore project), and the Manual for Civilization library.

MORE THAN MEETS THE EYE

Sherlock Holmes, the patron saint of detection, warned Watson that "there is nothing more deceptive than an obvious fact." Our new understanding of memory says facts, especially lots of them, can be misleading, distracting, and impede comprehension if used in the wrong context (or, is in the case of S., if you accumulate lots of them and never align them into coherent patterns or narratives). The point of gathering data is to paint a picture of the world as accurately as possible, with a precision that requires thinking slow, not fast. When so assembled, then the mind can use that picture of the world quickly and safely, in the daily business of life that relies on fast thinking, whether it is crossing the street or getting a first impression of someone at a reception. Facts find value and meaning only in the context of complex and dynamic systems.

Scientific disciplines from ecology to economics focus intently now on the study of interactions—chaos, complexity, and emergence. The neuroscience of memory has made great strides by applying reductionist techniques—investigating the smallest components of a system—to understand the basic elements of the brain such as dendrites, axons, and myelin. Given this rapid progress, as one practitioner said, the time has come "to link the molecular and cellular level with the systems level of analysis. This integration is the major challenge facing the science of memory, and might require, in addition to new methodologies, a change of *zeitgeist* or an amalgamation of approaches."

The physicist Robert Laughlin claims that the physical sciences have also stepped firmly from reductionist techniques to systems analysis. "This shift is usually described in the popular press as the transition from the age of physics to the age of biology, but that is not quite right. What we are seeing is a transformational world view in which the objective of understanding nature by breaking it down into ever smaller parts is supplanted by the objective of understanding how nature organizes itself."

The next advance in knowledge, in other words, will not just study elements comprising the agents of change, but change itself, the processes that combine to create complex behaviors and phenomena. To study climate change, for example, many fields of expertise come together to understand the interactions between atmosphere, ocean, fire, ice, biological cycles of energy production and consumption. The study of how Nature and its creatures organize themselves and change over time demands historical data, and lots of it. Like Sumerian scribes of yore, the researchers are gathering voluminous amounts of documentation in the course of doing business. Much of its value lies in its reuse over time. That is why the major funders of science in the United States, the National Science Foundation and the National Institutes of Health, now mandate that the researchers they support curate, preserve, and share their data with others.

Search engines, whose business is built from the ground up on the reuse of other people's data, also stake their future on managing and preserving the data they harvest. Google manages a lot of data; how much data it manages is proprietary information. But it is public knowledge that as of 2013 the company spent twenty-one billion dollars on data centers that process as well as store data for use. The cost of keeping data in storage is dependent on the cost of energy. It takes a lot of machines, generating a lot of heat and requiring industrial-strength air-conditioning systems, to keep our data planet spinning. The technology companies that monopolize search, social networks, and shopping are best positioned to develop the technologies that we need for long-term storage not because they are the first to understand the value of reusing data, but because they are best capitalized to work on new technologies. The collective memory systems that depend on machines for recording, preserving, and making sense of digital data are being developed at centers like these, under conditions of commercial secrecy. Together with the national security industry, these private enterprises are best positioned to sustain a massive if unknown percentage of our collective memory. They are also the ones developing technologies to extract more information from existing sources.

In the digital era, the key economic asset companies will compete for is not our labor, but our data. For both data-intensive industries and national defense, their primary assets are the data they get from us. Commercial companies gather this information more or less with our consent, although the average user clicking through license agreements cannot be said to be giving truly "informed consent." As for national security, Edward Snowden's revelations have put citizens across the globe on notice that our data are collected and used without our knowledge or consent. Both control vast amounts of intellectual and financial capital critical to solving a wired world's long-term information challenges. The fact that neither private commerce nor national security operate in ways transparent or directly accountable to the public is itself a significant risk factor for the future of our collective digital memory. That said, a growing number of critics question the propriety of companies' use of our data, and even more decry governments' abuse of our privacy and trust in the pursuit of national defense. In the coming decade, these two issues will be increasingly contested in the public policy arena, becoming key issues in elections and some moving on to the courts for final decision about how to interpret new and existing laws that touch on the definition of privacy and the ownership of data (including copyright and licensing issues).

Then, when the critical operating rules for the rights surrounding data have been defined and normalized, we will see another wave of social and technical innovation around the use of data.

(We will probably be using the word "sociotechnical" at this point, as we will be used to seeing technical innovations such as mobile computing change behavior, and behavioral demands spurring technical response.) These innovations will offer new values for old data (old in the digital realm being as young as two software or operating systems ago). Then demand—public and private—for access to secure and reliable digital content will balloon. Sustained long-term major investment in digital infrastructure will have proven itself the critical difference between societies that flourish in this world and those that flounder, fall behind, and as fail to protect their citizens during cyber warfare.

WHY OUR MACHINES NEED US

Now that we have discovered through empirical science that memory is a dynamic process, strongly inflected by emotion and spatialized in the brain, we have almost caught up with the ancient Greeks.

But we still lag in one area: the cultivation of knowledge for its own sake, above and beyond its usefulness in the manipulation of Nature or ability to return financial rewards on investment. The Greeks saw imagination as a divine dispensation from ignorance, a gift from the goddess of memory. Western science is the offspring of the marriage between the Greek love of knowledge for its own sake and the Christian view that knowledge of creation brings us knowledge of the creator. The intersection between the two is beauty and goodness. What is beautiful and good can be seen both as an end in itself and as a revelation of divine providence.

Over time, as we cracked the code of Nature, the providential usefulness of knowledge began to eclipse its beauty. Take Jefferson, quintessential man of science and technophile. In his view, the imaginative arts were to provide pleasure—though not too much. He complained to a friend that "the great obstacle to good education is the inordinate passion prevalent for novels, and the time lost in that reading which should be instructively employed . . . the result is a bloated imagination, sickly judgment, and disgust towards all the real business of life." The classification of Imagination or the Fine Arts was the smallest of the three categories in his library, a mere 20 percent of his books. A cultural bias toward instrumental knowledge continued to be a hallowed tradition in the United States, almost bred in the bone of the nation's political leaders. Even Abraham Lincoln, whose moral imagination was critical to the emancipation of slaves during the Civil War, was a technophile. As he told the Wisconsin State Agricultural Society in 1859, "I know of nothing so pleasant to the mind, as the discovery of anything which is at once new and valuable."

The Internet is extremely useful. At the same time it encourages the pursuit of curiosity for its own sake and democratizes it. Rather than condemning time spent pursuing our curiosity—online and elsewhere—as wasted hours of productivity, the pursuit of curiosity-driven questions should be encouraged with a clear conscience. Cultivating activities that are rewarding in and of themselves actually deepens our imagination. And—at the risk of making an instrumental argument—piquing our curiosity slows us down and allows for different modes of thought and processing information. Creativity flows from the suspended state of mind in which knowledge is nothing and attention is everything.

So how are we to think about thinking with our machines? We are advancing rapidly into a world where artificial intelligence creates, manages, and uses digital data for many tasks that humans are used to thinking only humans can do. Machines can replicate logical functions of human intelligence, and do it faster and better than we can. With good programming, they extrapolate patterns from masses of seemingly inchoate information, make astute predictions about our preferences, find the shortest route from A to Z, figure out betting odds more accurately than bookies, keep better time and know our schedules better than we do.

The realm of emotional intelligence, empathy, and imagination—all necessary for judgment in the context of incomplete information or conflicting aims—is beyond the reach of our machines. In the coming decades, we will take advantage of outsourcing logical tasks to our machines to free up time for more imaginative pursuits. As we teach our children how to use digital machines, we need simultaneously to cultivate their capacity for empathy and emotional intelligence by integrating the arts—the making of beautiful things and the telling of meaningful stories—into the new curriculum for digital literacy. Our machines will not grow a moral imagination anytime soon. They must rely on ours.

THE PERILS OF PERFECTIBILITY

The expansion of collective memory benefits everyone. The most readily adaptable animal is the one with the largest repertoire of stored experience to call upon. The smaller our repertoire of experience, the more vulnerable we are. Any society that periodically purges its collective memory of old, obsolete, or unorthodox views puts itself directly in harm's way. In the last century we were plagued by totalitarian regimes using massive distortion and erasure of people's histories in their audaciously criminal efforts to steer the future their way. This century is witness to the tyranny of theological terrorists who impose collective amnesia on their subjects by destroying historical and religious sites as ostentatiously as they kill their enemies.

But liberal democracies also imperil collective memory, though in more subtle ways. They tend toward an unnuanced faith in progress through science, "reason in action," which is a nearly invisible but still potent inheritance from the Enlightenment. Thomas Jefferson thought the human mind "is perfectible to a degree of which we cannot as yet form any conception." Since Jefferson's death, his beloved science made spectacular progress in decoding the language of Nature. Technology kept pace, applying the knowledge with ingenuity and alacrity, just as Jefferson hoped it would. But the impressive successes in these arenas have resulted in an overreliance on science as a model of knowledge not only for scientific matters, but also for countless things that science does not claim to address. As Sheila Jasanoff, an expert on scientific and

technological expertise, notes, "Science fixes our attention on the
knowable, leading to an over-dependence on fact-finding. Even
when scientists recognize the limits of their own inquiries, as they
often do, the policy world, implicitly encouraged by scientists, asks
for more research." The pursuit of perfect knowledge to solve
complex problems, she goes on, is pointless. "Uncertainty,
ignorance and indeterminacy are always present." Not every social
problem has a technological solution.

This secular faith in progress has its origins in a Christian
conception of history—the cosmic drama of fall from grace and
salvation through divine intervention. Though now wholly
divorced from any theological underpinnings, a residual evangel-
ical energy still emanates from the warm body of secularized faith,
promising miraculous technological interventions that can cheat
death, if not someday even defeat it. But as the physicist Steven
Weinberg notes, "Science addresses what is true, not what makes
us happy or good." In the wake of Darwin's discoveries of evolution
and its processes, the physical and life sciences moved decisively
away from a model of human perfectibility. In Nature, there is no
end, no consummation, no point of rest. On the contrary, the
work of life is endless adaptation, not incremental perfection. As
the political philosopher Isaiah Berlin said, "The historical process
has no 'culmination.' Human beings have invented this notion
only because they cannot face the possibility of an endless conflict."

Biology offers a word of caution about putting faith in *models of
perfect knowledge*. Long ago, some creatures inherited mutations
that made appendages suitable not just for navigating water, but
also, potentially, as limbs that could propel a creature on land or
wings that could keep it aloft in the air. Birds adapted to the air by
flying, but they were able to fly because they carried a useless
feature that under the right circumstances became useful: an
appendage that could become a winglike thing, a proto-wing.
Likewise, mammals did not evolve legs so that they could walk.
Mammals can walk because they inherited some preadapted
appendages that could become leglike things. Who is to say which
biological or cultural maladaptation will emerge in the future as a

key feature of success? In recent times, for example, new economies have shifted so radically from physical to intellectual productivity that people who were previously advantaged because of their physical prowess are now routinely sidelined by those with mental prowess—the triumph of the nerds.

Keeping evidence from past paradigms of knowledge, even those long since discredited, is the cultural equivalent of carrying maladaptations in the genetic code. Years, centuries, millennia can pass. But when our environment suddenly changes, as it is changing now, it may be the odd trait, tradition, or idea we carry by accident that helps us to adapt. Then those maladaptations may turn out to be useful. They become exaptations, inherited features recruited for purposes other than originally selected for. The logic of adaptability means the greater the diversity of traits a creature carries, the greater the chance it has to adapt to changing circumstances. As the rate of biological species' extinctions accelerates, we are learning just how important diversity is in the biological world.

You can imagine the resurrection of classical writers in the Renaissance and their eager adoption by Europeans as a cultural exaptation of sorts. The writings of Stoic and Epicurean philosophers became alive again in the midst of Christian culture in upheaval. They were being used by eager readers in religious and political debates that were wholly alien to the classical world from which they sprung. And yet for the previous thousand years their books were judged to be obsolete, ignorant, even pernicious. They were destroyed or allowed to decay and disappear as the cultures of Christianity and Islam, struggling for survival and dominance, became monocultures. One orthodoxy came to dominate the information landscape by eradicating the natural diversity of beliefs and perspectives. But the texts that nonetheless survived in Byzantine and Islamic centers of learning acquired radically new values with the passage of time.

As the world globalizes and shrinks, so will the world of ideas. A monoculture of ideas renders us as vulnerable to catastrophic loss and failure as a monoculture in the agricultural world. We want more knowledge, not less, about the multiple ways to be human. We cannot know what the future value of any archaic or seemingly irrelevant body of knowledge may be. Our obligation to future generations is to ensure that they can decide for themselves what is valuable.

WHEN WE ARE NO MORE

We are a fortunate species. We see the world not as it is, but as our minds assemble it from memory. We create a richly detailed 4-D diorama of the world that guides our every action. The model changes as we do and as the world does. As we add new details that deepen our understanding of ourselves and others, some memories are modified and strengthened, while others recede in their vividness.

We are fortunate because we can so readily appropriate the memory of others to aggrandize our mental model with knowledge of worlds altogether foreign to our experience. Private and public, autobiographical and historical, real and imaginary—the whole memory of humanity can claim its place in our mental model. The richer the model, the quicker and keener our grasp of the novel, the faster we recognize the news that does not quite fit into our understanding of the world. We can infer the significance and meaning of what we perceive by finding a categorical or emotional match in our mind's archive, an archive enriched by the memory of humanity.

From the time our ancestors first discovered that objects could extend the reach of their thoughts and feelings, we began to accelerate our evolution through self-cultivation. Curiosity and awareness of our mortality, together with our ability to send knowledge forward into the future, have culminated in something truly new. We have gained dominion over land, sea, and skies, remaking Nature in our image. We optimize the biochemistry of plants and animals for our consumption, regulate our bodies and minds with pharmaceuticals, travel across time zones without regard for the circadian rhythms of our bodies. Our voices resound instantaneously across time and space on digital networks. Humanity is, as Milosz says, an elemental force conscious of transcending Nature, for we live by memory of ourselves, that is, in history. We are close to reversing what has long been a defining distinction between humans and all other life. Until now, we have been the sole species able to use nongenetic means to transmit information across time and space. Outsourcing memory has been our strategic advantage for at least forty thousand years. But now, with the ability to alter genetic material and potentially turn that, too, into a medium for human information, we may embed DNA itself with our acquired knowledge and send it forth into the future.

We have sprung this new world of digital data on ourselves without advance warning. We are now several decades into this uncontrolled (and uncontrollable) experiment, and have yet to catch our breath. We are moving, pushmi-pullyu-like, in opposing directions—quickly adapting and domesticating the digital world, at the same time expanding into unknown territories. The faster we move, the less predictable our path becomes. In 1997, when I saw that we will not have libraries and archives full of hard-copy "rough drafts" of present-day history, it seemed we could not adapt quickly enough to avoid the loss or corruption of the past. But since 1997, the power of our machines to extrapolate a wealth of information from even fragments of the past—from bird specimens and glass plate negatives to broken lacquer discs and ships' logs— tells a different story. We are beginning to learn how much we can afford to lose and still come to know our own history.

Today, we see books as natural facts. We do not see them as memory machines with lives of their own, though that is exactly what they are. As soon we began to print our thoughts in those hard-copy memory machines, they began circulating and pursuing their own destinies. Over time we learned how to manage them, share them, and ensure they carried humanity's conversations to future generations. We can develop the same skills to manage and take responsibility for digital memory machines so that they too outlive us and, like books, "derive from people, but also from radiance, heights." Whether we do or not is now in our hands.

ACKNOWLEDGMENTS

This book was born, like many, in the Library of Congress, and my first and deepest thanks go to Winston Tabb, librarian and friend extraordinaire. He brought me in to the library and gave me license to follow my curiosity where it took me. Above and beyond sharing his profound knowledge of libraries, copyright law and practices, collectors, the Congress, and anything I asked about, he set an example of the finest public service imaginable, when getting the right result always takes precedence over getting credit for doing so. And when I left the library, he gave me his personal copy of Millicent Sowerby's work on Jefferson's library, the seed from which this book sprouted.

Other colleagues at the library, in particular Sam Brylawski, Michael Grunberger, John Y. Cole, and the late Peter Van Wingen, generously and congenially shared their knowledge with me over the years. I am indebted also to two outstanding library leaders: Deanna Marcum, for her uncompromising commitment to the preservation of and ongoing access to our collective memory to advance knowledge, and her steadfast encouragement to advance my own; and Michael Keller, for his fearless dedication to the future of libraries in the digital age. My thanks also to his colleagues, the superb professionals at Stanford University Libraries.

I have been helped and inspired by Carl Haber of the Lawrence Berkeley National Laboratory; and Brewster Kahle and Wendy Hanamura at the Internet Archive.

In the making of this book, my thanks go to the incomparable Ike Williams and Katherine Flynn, to Peter Ginna, and to Rob Galloway and George Gibson. Thanks also to Kate Wittenberg, Megan Prelinger, and Cheryl Hurley, who kindly gave me advice when I most needed it—often before I knew I needed it. The responsibility for any and all errors of fact, judgment, and interpretation here is mine alone.

Above all, my thanks go to David Rumsey, whose map library is the greatest universe of spatial memory—both artifactual and digital—in the world, and who works unstintingly to grow it and make it free to all. He keeps my spirits high and my feet on the ground.

NOTES

CHAPTER ONE: MEMORY ON DISPLAY

6   *an average of forty-four days*: *Preserving Our Digital Heritage*, vol. 2, 53.

6–7   *Square Kilometre Array*: Drake, "Cloud computing," 543. For more information about the Square Kilometre Array (SKA), see skatelescope. org. Construction of the SKA is scheduled to begin in 2018. When complete in 2030, it will deploy one square kilometer of collecting space with antennae in both Australia and South Africa.

7   *from 2.7 billion terabytes to 8 billion*: http://cdn.idc.com/research/ Predictions12/Main/downloads/IDCTOP10Predictions2012.pdf. Accessed June 8, 2012. (Although this IDC projection was widely cited at the time, this link is no longer live.) "A *petabyte* is the equivalent of 1,000 terabytes, or a quadrillion bytes. One *terabyte* is a thousand gigabytes. One *gigabyte* is made up of a thousand *megabytes*. There are a thousand thousand—i.e., a million—petabytes in a *zettabyte*." Jonathan Shaw, "Why 'Big Data' is a Big Deal." *Harvard Magazine*, March-April 2014, 33.

CHAPTER TWO: HOW CURIOSITY CREATED CULTURE

15   *a collective form of memory*: Many biologists identify imitative and imprinted behaviors among animals as cultural features—chicks that learn bird songs from other birds, chimps that imitate other chimps using sticks to dig up grubs, and so on. The focus here is not on learning through direct physical imitation, but acquiring knowledge from others *indirectly* through physical objects that carry information.

15   *DNA extraction and analysis*: On recent findings in genetic research into the origins of *Homo sapiens*, see Pääbo, *Neanderthal Man*; and Tattersall, *Masters*.

16   *distinctively human behaviors*: On disagreement about how to interpret the evidence that argues for attributing these behaviors to Neanderthals, see Appenzeller, "Old Masters."

16   *long-distance exchange of goods*: On objects becoming constituents of the "web of human relations," see Shryock and Smail, "History and the 'Pre'," 724.

16   *thinking with things*: Some scientists postulate that toolmaking somehow catalyzed the development of language, and have mapped activity in the brain regions activated by making Paleolithic hand axes that overlap with some required by speech. That said, Neanderthals, who are thought to have made tools, did not have language. The causal relationship between genetic potential and cultural practice remains obscure. See Normile, "Experiments Probe Languages' Origins."

16   *occurrences of cave art*: Indonesian cave art that looks very much like the paintings in Eurasia has recently been dated to about forty thousand years ago. If these dates hold up, scientists will need to revise the prevailing view on when and how *Homo sapiens* came out of Africa and settled the globe. See Aubert et al., "Pleistocene cave art."

16   *early art objects*: Clottes, *Cave Art*, 11ff. The earliest possible art object, found in the Blombos Cave in South Africa, is a "worked and polished hematite stone, decorated with a complex engraved motif of three parallel lines and a series of crosshatches" found in a layer dated to around seventy-five thousand years ago. In 2009, researchers announced they had found thirteen engraved ochre pieces, some dating one hundred thousand years, and identified them as evidence of symbolic thinking. Balter, "Early Start." Anthropologist Andrew Shryock and historian Daniel Lord Smail point out that the modern human "cultural assemblage" appears, almost abruptly, in Europe around forty to fifty thousand years ago, but was likely assembled more slowly in Africa, whence modern *Homo sapiens* migrated. "It appears to burst out in Europe only because Africans, that is to say modern *Homo sapiens*, carried it with them when they migrated into Europe around 50,000 years ago." Shryock and Smail, "History and the 'Pre'," 716.

18   *"you may freely eat"*: Genesis, 1:15–17.

19   *libraries as thought experiments*: One such library is being assembled by the Long Now Foundation, an organization dedicated to encouraging long-term thinking. The projected library of thirty-five hundred volumes is called the Manual for Civilization. Its scope focuses on four categories: a cultural canon, the mechanics of civilization, rigorous science fiction, and long-term thinking, futurism, and relevant history. http://blog.longnow.org/02014/02/06/manual-for-civilization-begins/.

19 *the Western model of historical thinking*: See Smail, *On Deep History*, 12–39, on what he calls "the grip of sacred history."

21 *cuneiforms were created before 3300 B.C*: By 3200 B.C., Egyptians were using hieroglyphs.

23 *"Man is by nature a cultural animal"*: Bidney, "Human Nature," 376.

24 *our evolutionary niche as generalist*: Tattersall, *Masters*, 228.

25 *writing was invented multiple times*: "Cumulative evidence around the world suggests that writing was invented at least three times in the last part of the fourth millennium B.C.E., and at least three more times in different parts of the world in later periods." Wolf, *Proust and the Squid*, 47.

25 *over thirty thousand tablets survive*: http://www.britishmuseum.org/research/research_projects/all_current_projects/ashurbanipal_library_phase_1.aspx.

26 *"We, and our judgment"*: Montaigne, "Apology for Raymond Sebond," in *Complete Works* (hereafter cited as *CW*), 455.

28 *"The clothes I wear"*: Milosz, "On the Effects of the Natural Sciences," in *Visions*, 21.

CHAPTER THREE: WHAT THE GREEKS THOUGHT: FROM ACCOUNTING TO AESTHETICS

30 *housed three hundred thousand scrolls*: Battles, *Library*, 8.

30 *the "Great Vanishing" of classical memory*: Greenblatt, *The Swerve*, 86.

30–31 *Antony pillaged the Library of Pergamum*: Ibid., 281. Luciano Canfora, in his book *The Vanished Library*, draws different conclusions from the same scant evidence: that it was not the library but a dockside depot warehousing assorted cargo (including books) that burned in 48 B.C., and that the story of Antony expropriating books from Pergamum to give to Cleopatra was a "calumny . . . perhaps intended as a gibe at Antony's ignorance of literary matters." See pp. 69–72 and 91ff.

32–33 *experts on ancient slavery*: For a comparative perspective on slavery in the ancient world, see David Brion Davis, *Inhuman Bondage: The Rise and Fall of Slavery in the New World* (New York: Oxford University Press, 2006), chap. 2, "The Ancient Foundations of Modern Slavery."

35–36 *Alzheimer's disease attacks the hippocampus*: Hassabis et al., "Patients with Hippocampal Amnesia," 1726; Kahana et al., "Neural Activity," 1726.

36 *2014 Nobel Prize in Physiology or Medicine*: Underwood, "Brain's GPS," 149.

36 *Cicero on Simonides*: Cicero, *De Oratore*, cited in Yates, *The Art of Memory*, 17.

37   *palm as a mnemonic device*: See the exhibition catalog *Writing on Hands: Memory and Knowledge in Early Modern Europe*, organized in 2000 by the Trout Gallery, Dickinson College.

37–38   *memory reinforced through physical objects*: The special relationship we have with the objects we make is not quite as straightforward as it first appears. Why is it natural for us to make things to think with but not for others species? Some philosophers, espousing the "extended mind theory," propose that our consciousness essentially appropriates and colonizes the physical world as we come to know it. We create tools and technologies to extend our will into the world. This is different from the extension of self through ownership of property ("my land, my slave, my house, my cat"). The former is found universally among human cultures; the latter is highly variable among cultures. For an overview of the debate, see Clark, *Supersizing the Mind*.

38   *residential-care regimes for Alzheimer's patients*: Danziger, *Marking the Mind*, 271.

38   *physical movement and thought*: The enhancement specifically of creative thinking while walking versus sitting is reported in Oppezzo and Schwartz, "Give Your Ideas Some Legs."

39   *interactive timelines, maps, charts, and infographics*: The role of visualization in gisting information is the subject of "Learning to See Data" by Benedict Carey, *New York Times*, March 27, 2015.

39   *going to a library for access to information*: Library is a modern word, derived from the Latin word for book, *liber*. Other modern languages, such as French and Russian, take their word for library from the Greek for book or papyrus, *biblos* (*bibliothèque* and *biblioteka* respectively). Contemporary connotations of the word are broadening beyond books and even texts to mean any collection of content, such as "my music library" (typically in MP3 formats on a computer). This contrasts with "archives," used to denote a collection on a vastly larger scale.

40   *no useful archaeological evidence*: Battles, *Library*, 30.

42   *essential model for the library in the digital age*: In 2002, the Bibliotheca Alexandrina, a library, museum, and cultural center built near the site of the original library, opened on the site of the old Mouseion. It houses books and manuscripts relating to the contemporary and historic Mediterranean, many donated from the Bibliothèque nationale de France, along with extensive microfilm and digital collections of manuscripts and rare books.

42   *"For this invention will produce"*: *Phaedrus*, 275a–b. This is actually Plato quoting Socrates, who is quoting the Egyptian Thamus rebuking the god Theuth for inventing writing.

44   *muses inspire or prompt the performers*: Danziger, *Marking the Mind*, 28ff.

47   *legacy of classical Greece and Rome*: In Byzantium, both lay and clerical

elites were taught to read from Homer's *Iliad*. "From Homer they went on to the great dramas of Aeschylus, Sophocles, Euripides, and Aristophanes, and then to Demosthenes, Thucydides, Aristotle, and Porphyry." Carr, "Reading, Writing, and Books," 181; and for examples of the works they studied, see 192–99.

CHAPTER FOUR: WHERE DEAD PEOPLE TALK

48   *survived in complicated fragments*: Ancient texts seeped into southwestern Europe over the centuries through the Venetian Empire, with close ties to Byzantium; and in the West, through Moorish Spain, with close ties to an Islamic culture of learning.

49   *between 150 and 200 million books*: Febvre and Martin, *The Coming of the Book*, 186.

49   *Montaigne learned to read*: His father had somewhat eccentrically decided to raise his son to be fluent in Latin. Imposing his own form of immersion learning on little Michel, his father did not allow him either to speak or hear anything except Latin for the first decade of his life.

49   *two Bibles face each other*: On the Giant Bible of Mainz, see http://www.loc .gov/exhibits/bibles/the-giant-bible-of-mainz.html; and on the Gutenberg Bible, see http://www.loc.gov/exhibits/bibles/the-gutenberg-bible.html.

50   *he published and republished volumes*: The history of Montaigne's publishing is hard to untangle. He put out several editions in his lifetime, not only adding new essays but also going back to previously published ones and adding more to them. He published the first edition of his essays in 1580, and in 1582 he published a second. In 1588, he published the third edition, this one greatly expanded with new essays. He was preparing yet a fourth edition when he died in 1592. In 1598, a posthumous edition of his essays appeared, including many of his late changes and additions. All this has provided grist for literary historians over the ages. It is virtually impossible to pick up any copy of Montaigne's essays that does not include a lengthy introduction by the editor and translator explaining which editions of which essays they chose to include and why.

50   *Montaigne saw in himself what was universal*: "I set forth a humble and inglorious life; that does not matter. You can tie up all moral philosophy with a common and private life just as well as with a life of richer stuff. Each man bears the entire form of man's estate." Montaigne, "Of repentance," *CW*, 611.

50   *a new generation of memoirists*: In particular, memoirs about addiction, divorce, abuse, incest, and other topics that were seldom publicly discussed in the past proliferate today as a way of creating a contemporary

biographical narrative that allows readers to see themselves in a larger social context and make sense of their own journey.

51 *"In the year of Christ 1571"*: Montaigne, *CW*, ix–x.

51 *"The shape of my library is round"*: Frame, *Montaigne: A Biography*, 121; and Montaigne, "Three kinds of association," *CW*, 628–29.

52 *he thought he was dying*: He describes this incident in his essay, "Of practice," *CW*, 268–269.

52 *Montaigne repeatedly witnessed his father*: Bakewell, *How to Live*, 227.

52 *"the thing I fear the most is fear"*: Montaigne, "Of fear," *CW*, 53.

52–53 *"nature has lent us pain"*: Montaigne, "Of experience," *CW*, 837–38.

53–54 *"my relatives and friends"*: Montaigne, "To the Reader," *CW*, 2.

55 *overrun by warring factions*: Montaigne was forced on one occasion to evacuate his household and seek safety farther afield. But fear of the plague shut many doors tight. "I, who am so hospitable, had a great deal of trouble finding a retreat for my family: a family astray, a source of fear to their friends and themselves, and of horror wherever they sought to settle, having to shift their abode as soon as one of the group began to feel pain in the end of his finger." Montaigne, "Of physiognomy," *CW*, 801–2.

55 *assassination as political tool*: Henri III had the Duke of Guise assassinated in December 1588. Then the following August, he was assassinated.

56 *reforming Christian faith*: See MacCulloch, *The Reformation*, for a comprehensive look at current scholarship on late medieval religion as well as the Reformation.

56 *Martin Luther's "Ninety-Five Theses"*: On printing and religious propaganda, see Febvre and Martin, *The Coming of the Book*, 287–319. Especially powerful was the new availability of scripture in local languages. "Between 1466 and 1522 there were twenty-two editions of the Bible in High or Low German; it reached Italian in 1471, Dutch in 1477, Spanish in 1478, Czech around the same time, and Catalan in 1492. In 1473–4 French publishers opened up a market in abridged bibles, concentrating on the exciting stories and leaving out the more knotty doctrinal passages." (MacCulloch, 73).

60 *"I have known books"*: Montaigne, "Of physiognomy," *CW*, 808.

60 *"My ignorance will excuse me"*: Montaigne, "Of books," *CW*, 301.

61 *"I aim [in my essays]"*: Montaigne, "Of the education of children," *CW*, 109.

62 *"there is no desire more natural"*: Montaigne, "Of experience," *CW*, 815.

CHAPTER FIVE: THE DREAM OF THE UNIVERSAL LIBRARY

64–65 *"fugitive fermentation of an individual brain"*: Thomas Jefferson to Isaac McPherson, August 13, 1813.

66 *"Enlighten the people generally"*: Jefferson in a letter to P. S. Dupont de Nemours, April 24, 1816.

68 *1,256 volumes at Monticello*: *The Jefferson Bicentennial 1743–1943*, 25.

68 *a self-confessed bibliomaniac*: "Sensible that I labour grievously under the malady of Bibliomanie, I submit to the rule of buying only at reasonable prices, as to a regimen necessary in the disease." Jefferson in a letter to Lucy Ludwell Paradise, June 1, 1879. http://founders.archives.gov/documents/Jefferson/01-15-02-0166.

68 *"While residing in Paris"*: Jefferson in a letter to Samuel H. Smith, September 21, 1814.

70 *books burned in Capitol fire*: Before the fire, the library had 3,076 volumes as well as maps, charts, newspapers, and congressional records. *The 1812 Catalogue*.

70 *"Architecture is my delight"*: The quote is attributed to Jefferson by Margaret Bayard Smith, who visited him in 1809. Giordano, *The Architectural Ideology*, 150.

71 *"Those who opposed the bill"*: *Annals of Congress*, 13th Cong., 3rd sess., 28: 1105–6.

72 *"I do not know"*: Jefferson in a letter to Samuel H. Smith, September 21, 1814. In 1802, while serving as president, Jefferson drew up a list of books the library should have, necessary "to the deliberations of the members as statesmen, and . . . omitted those desirable books, ancient and modern, which gentlemen generally have in their private libraries, but which cannot properly claim a place in a collection made merely for the purposes of reference." Like Montaigne, Jefferson was never one to be hobbled by consistency. In any event, despite President Jefferson's insistence that "books of entertainment" were out of scope, by 1812 there were many books in the categories of "Poetry and the Drama, Works of Fiction, Wit, &c" and "Arts and Sciences, and Miscellaneous Literature." These were usually donated by members for use by their peers, so they ended up with a collection not unlike that of a gentleman's club in London. *The 1812 Catalogue*, xiii ff.; and Goodrum and Dalrymple, *The Library of Congress*, 12.

72 *Ossian admired by Jefferson*: Scottish poet James Macpherson published a series of epic poems attributed to Ossian. He claimed they were translations of ancient Scots Gaelic epics that had been transmitted orally over generations. Ossian was thought to be the Gaelic Homer.

73 *"It is the duty of every good citizen"*: Jefferson in a letter to Hugh P. Taylor, October 4, 1823.

73 *"I endeavor to collect"*: Jefferson in a letter to James Madison, January 12, 1789.

74 *"the worthiest minds, who lived in the best ages"*: Montaigne, "Of the education of children," *CW*, 115.

74 *"universal toryism"*: Jefferson in a letter to John Norvell, June 11, 1807, cited in Sowerby, I: 157.

75 *Benjamin Franklin's marginalia: American Treasures in the Library of Congress*, 68–69.

75-76 *"The earth belongs to the living"*: Jefferson in a letter to James Madison, September 6, 1789.

76 *"I cannot live without books"*: Jefferson in a letter to John Adams, June 10, 1816.

77 *"I feel a much greater interest"*: Jefferson in a letter to Nathaniel Macon, January 12, 1819.

79 *Daniel Kahneman has pointed out*: Kahneman, *Thinking*, 417–18.

79 *collect and curate artifacts*: The words "curiosity" and "curation" share the same etymological root, *cura*, or care. According to the *Oxford English Dictionary*, in the eighteenth century curiosity meant carefulness, scrupulousness, accuracy as well as inquisitiveness and the desire to know.

81 *Packard Campus for Audio-Visual Conservation*: www.loc.gov/avconservation/packard/.

CHAPTER SIX: MATERIALISM: THE WORLD IS VERY OLD AND KNOWS EVERYTHING

85 *"the tranquil pursuits of science"*: "Within a few days I retire to my family, my books and farms ... Never did a prisoner, released from his chains, feel such relief as I shall on shaking off the shackles of power." Jefferson in a letter to Pierre Samuel Dupont de Nemours, March 2, 1809.

86 *"The word 'science'"*: Shapin, *The Scientific Revolution*, 5–6, n3.

88 *species ever go extinct*: See Martin, *Thomas Jefferson*, 96–114; and Thomson, *Jefferson's Shadow*, 86–97.

88 *"reason in action"*: "In an age in which reason was venerated, science was esteemed as the intellectual manifestation of human reason in action." Cohen, *Science and the Founding Fathers*, 60.

88 *"Freedom [is] the first-born daughter"*: Jefferson in a letter to François D'Ivernois, February 6, 1795.

88–89 *American Philosophical Society*: For the early years of the American Philosophical Society, see Lyons, *The Society for Useful Knowledge*.

89 *"have spent the prime of our lives"*: Italics in original. Jefferson in a letter to Harvard president Joseph Willard, March 24, 1789, written from Paris just months before the storming of the Bastille.

89 *"Science is important to the preservation of our republican government"*: Quoted in Martin, *Thomas Jefferson*, 56.

90 *exotic live specimens*: Ibid., 14; and Thomson, *Jefferson's Shadow*, 224ff.

90   *Jefferson's admiration for Francis Bacon*: Jefferson in a letter to John Trumbull, February 15, 1789. He considered Francis Bacon to be one of the three greatest men ever to have lived (the other demigods being Isaac Newton and John Locke). Bacon's scheme, widely known by subsequent generations, was adopted by Diderot for his *Encyclopédie méthodique*. A diagram of this system of human knowledge appeared as a frontispiece in the first volume of the encyclopedia, a book well thumbed by Thomas Jefferson.

91   *"Lord Bacon founded his first great division"*: Jefferson in a letter to A. B. Woodward, March 24, 1824.

91–92   *"In classing a small library"*: Jefferson's annotation to his catalog of March 1783 can be viewed at: http://www.masshist.org/thomas-jeffersonpapers/.

92   *letter to A. B. Woodward*: Jefferson in a letter to A. B. Woodward, March 24, 1824.

92   *Jesus was a materialist*: "Indeed Jesus himself, the founder of our religion, was unquestionably a materialist as to man. In all his doctrines of the resurrection he teaches expressly that the body is to rise in substance. In the Apostles' Creed we all declare that we believe in the 'resurrection of the body.'"

93   *"Metaphysics have been incorporated with Ethics"*: Composed sometime between 1820 and 1825 as a preface to "A Catalogue of Books Forming the Body of a Library for the University Of Virginia" and published in the *University of Virginia Alumni Bulletin*, November 8, 1895, 79. Thanks to Heather Riser of the Albert and Shirley Small Special Collections Library at the University of Virginia for this reference.

94   *historical eras could be dated relative to each other*: There was nothing here that contradicted the divine creation. The argument was about when and how God had created the world.

96   *"the God whom science recognizes"*: James, *The Varieties of Religious Experience*, 536.

97   *roots of materialism in Christianity*: Christianity absorbed a good deal of Greek and Roman pagan thought in its formative years. But it is not clear that Christianity felt the influence of pre-Socratic thinkers, such as Empedocles, nor that of the Epicureans. The poem that propounds materialism, for example, *On the Nature of Things* by the Epicurean poet of the first century B.C., Lucretius, was lost to Western thought until 1417. See Greenblatt, *The Swerve*.

100   *"Evidence of the antiquity of man"*: From J. W. Burrow; and note David Friedrich Strauss published his book *The Life of Jesus, Critically Examined* in 1835 to 1836, which was translated into English by George Eliot in 1846.

101   *natural philosophy and natural history collapse into science*: In French, German, and Russian, the general term for science (*la science, Wissenschaft,*

and *nauka* respectively) covers all disciplines, including what in English are called the humanities and social sciences.

101 *British Association for the Advancement of Science*: This was in stark contrast to the Royal Society of London for Improving Natural Knowledge in London, chartered by King Charles II in 1662. It had lost considerable scientific luster since the days when it counted Isaac Newton, Robert Boyle, Robert Hooke, and Edmond Halley as members. See Danielson, "Scientist's birthright," 1031. See also *OED*, 1834; Ross, "Scientist: the story of a word"; and Snyder, *The Philosophical Breakfast Club*, 2–3.

101 *"scientist" came only slowly into common parlance*: Google Ngram search for "scientist" (case insensitive) shows the word essentially not in use at all until the 1860s; Snyder, *The Philosophical Breakfast Club*, 297–98. People resisted the term as a barbarism. It caught on in the United States faster; as Snyder (361) says, Americans are "always more open to new things. Indeed, the term became closely associated with American scientists, and by 1874 its English roots were forgotten, the president of the Philological Society in England referring to 'scientist' as 'an American barbarous trisyllable.'" The Ngram Viewer, incidentally, is in early stages of development. Its Google Books database comprises titles held chiefly in U.S. academic libraries, one particular slice of the digital universe. The corpus is text from roughly 7.5 million books, about 6 percent of all books ever published; it includes no journals, newspapers, or unpublished materials such as manuscripts, letters, and so forth. The Ngram is fun to use and quite suggestive, but like all searches of large databases, can only tell us what the selected data know. In this case at least what it tells us accords with what historians have deduced from research using other methods and sources. See Natasha Singer, "In a Scoreboard of Words, a Cultural Guide," *New York Times*, December 7, 2013.

102–103 *invention of image capture*: The French inventor Joseph Nicéphore Niépce is believed to have created the first image using a photographic process in 1826. On the 1860s "phonautogram," see www.firstsounds. org/sounds/. By 1878, Edison filed for a patent on a phonograph, a cylinder wrapped in tinfoil that could be encoded for audio waves. See www. aes.org/aeshc/docs/recording.technology.history/notes.html. Thanks to Sam Brylawski for this and other references about sound recording.

104 *giddy pace of the Red Queen and Alice*: "Now, *here*, you see, it takes all the running you can do, to keep in the same place," the Red Queen explains in *Through the Looking-Glass and What Alice Found There*.

105 *"You say that we go round the sun"*: Doyle, "A Study in Scarlet," chap. 2, in *The Adventures of Sherlock Holmes*.

105 *"the objects of natural knowledge"*: Shapin, *The Scientific Revolution*, 162.

106 *"The most powerful storehouse of value"*: Ibid., 164.

CHAPTER SEVEN: THE SCIENCE OF MEMORY AND THE ART OF
FORGETTING

109 *biology is neither personal nor cultural destiny:* Much of what we know
about human memory is based on knowledge gleaned from two sources:
humans who have memory disorders and experiments conducted on
other species. Animal studies usually involve genetic tinkering and other
means of creating memory disorders in lab animals, procedures not
permissible to perform on humans. Such studies rely on the vital, if qual-
ified, identification of humans with other forms of life. One of the pressing challenges in life sciences is to develop a keener understanding of
when models based on rats, mice, or sea slugs work in humans and when
they do not. The biologist Richard Lewontin cautioned strenuously
against using biological models of evolution to explain cultural practices
and how they change. "We would be much more likely to reach a correct
theory of cultural change if the attempt to understand the history of
human institutions on the cheap, by making analogies with organic
evolution, are abandoned. What we need instead is that much more
difficult effort to construct a theory of historical causation that flows
directly from the phenomena to be explained." Lewontin, "The Wars
over Evolution," 54. He was addressing primarily his fellow natural
scientists and social scientists who use evolutionary psychology to
"explain" human behaviors and cultural practices. Several disciplinary
perspectives on uses of biology in history and history in biology are
published in the *American Historical Review*, "*AHR* Roundtable: History
and Biology," 119: 1492–629.

110 *prepares a miniature map of its environs:* Gazzaniga, *The Mind's Past,* 74–
75. "We have predictive perception . . . what we see is not what is on the
retina at a given instant, but is a prediction of what will be there. Some
system in the brain takes old facts and makes predictions as if our perceptual system were really a virtual and continuous movie in our mind."

110 *we spatialize that information:* Research suggests that place is measured
by time in the brain. "Essential to spatiotemporal encoding are 'place
cells' [in the hippocampus] that fire when an animal passes through a
precise position in its environment and theta oscillations, brain impulses
that act as an internal 4–10 Hz clock . . . Thus time in the hippocampus
is organized topographically, resembling a series of local time zones. The
findings imply that, at any given time, the hippocampus encodes an
extended segment of the environment—not just a point in space." "The
map in your head," *Nature* 459: 477.

110 *"By virtue of the unconscious status":* Kandel and Squire, "Neuroscience."

111 *modern biology as a science of information:* "Modern biology is a science of
information," editorial, David Baltimore, *New York Times,* June 25,

2000. By the 1940s, fundamental advances in biochemistry had already transformed the life sciences "from a discipline concerned with enzymes and the transformation of energy (that is, with how energy is produced and utilized in the cell) to a discipline concerned with the transformation of information (how information is copied, transmitted, and modified within the cell)." Kandel, *In Search of Memory*, 242.

111 *"The new genome"*: Pollack, *The Missing Moment*, 14.

112 *crucial processing steps*: See Yang et al., "Sleep promotes."

112 *poems are "like dreams"*: "When We Dead Awaken: Writing as Re-Vision," in *On Lies, Secrets and Silence. Selected Prose 1966–1978*. New York: W. W. Norton, 1979.

112–113 *"Biological value lies only in learning"*: Quoted in Jonathan Weiner, *Time, Love, Memory: A Great Biologist and His Quest for the Origins of Behavior* (New York: Alfred A. Knopf, 1999), 138.

114 *"The retrieval of the consolidated memory"*: Kandel and Squire, "Neuroscience." "Reactivated memory undergoes a rebuilding process that depends on de novo protein synthesis. This suggests that retrieval is dynamic and serves to incorporate new information into pre-existing memories. However, little is known about whether or not protein degradation is involved in the reorganization of retrieved memory . . . It also provides strong evidence for the existence of reorganization processes whereby pre-existing memory is disrupted by protein degradation and updated memory as reconsolidated by protein synthesis." Lee et al., "Synaptic Protein Degradation," 1253.

114 *"the progressive post-encoding stabilization of the memory trace"*: *Science of Memory*, 165.

114 *memory is most vulnerable*: "During that phase, but not afterwards, the memory item is susceptible to amnestic agents." Ibid., 165.

116 *analog and digital circuits*: O'Reilly, "Biologically Based Computational Models"; and Eisenberg, "What's Next."

117 *"It is clear that the brain is much more"*: O'Reilly "Biologically Based Computational Models," 94.

118 *emotions are associated with our senses*: Dolan, "Emotion, Cognition, and Behavior."

118 *"Pure emotion can be viewed"*: Greenfield, *The Private Life of the Brain*, 21.

118 *"without requiring any conscious memory content"*: Kandel and Squire, "Neuroscience."

118 *directed attention requires effort*: True, we can make choices about where to direct our thoughts. We can decide to focus on what the woman behind the counter is saying in response to our question; we can choose between red wine versus white with pork or among various cable subscription plans, including none at all. We can direct our thoughts to coordinating muscles and synchronizing our breathing as we practice our

backhand stroke. We can memorize a tune, a poem, or a string of numbers. But that concentration, no matter how great the effort to sustain, represents a small fraction of the brain's activities. If, during any of these acts of forced concentration and exercise of working memory, we were to hear a loud noise, feel a wave of concussed air, and smell something acrid, we would immediately turn our attention to figuring what caused the noise, locate where it came from, and scan our environs for an escape route from the fire.

119 *"Emotion exerts a powerful influence on reason"*: Dolan sees emotion as "states that index occurrences of value." See "Emotion, Cognition, and Behavior," 1191.

119 *empathy and fellow feeling*: Pronin, "How We See Ourselves and How We See Others."

121 *as he lay still under the covers*: Luria, *Mnemonist*, 151.

121 *the word for beetle,* zhuk: Ibid., 83–84. He went on to mention a lump of coal, the splash made when his grandmother was pouring tea, unmelted tallow, and so on.

121 *the sheer density of sensory associations*: The proliferation of unedited, unconsolidated associations bears similarity to certain features of autism, see p. 128.

122 *"convert encounters with the particular"*: Jerome S. Bruner, foreword to the first edition of Luria, *Mnemonist*, xxii.

122 *"This makes for a tremendous amount of conflict"*: Luria, *Mnemonist*, 116.

122 *"to weigh one's words"*: Ibid., 119. "And take the expression *to weigh one's words.* Now how can you weigh words? When I hear the word *weigh,* I see a large scale—like the one we had in Rezhitsa in our shop, where they put the bread on one side and the weight on the other. The arrow shifts to one side, then stops in the middle . . . But what do you have here—to *weigh one's words!*" In the original Russian text, the examples I use for "arm" are given as uses of the word *ruchka* to mean "child's arm," "door handle," or "penholder."

122 *he had a hard time remembering faces*: Ibid., 64.

123 *"At one point I studied the stock market"*: Ibid., 157.

123 *"I was passive for the most part"*: Ibid., 157.

124 *"living pasts and dead pasts"*: Le Corbusier, "When the Cathedrals Were White: Journey to the Country of Timid People." quoted in *New York Times Magazine*, November 11, 2001, p. 23.

CHAPTER EIGHT: IMAGINATION: MEMORY IN THE FUTURE TENSE

126 *"How do our minds get so much"*: Tenenbaum et al., "How to Grow a Mind," 1279.

126–127   *"Great is this power of memory"*: Augustine, *Confessions*, 197.
128   *cognitive disorders such as autism*: Guomei Tang et al., "Loss of mTOR-Dependent Macroautophagy Causes Autistic-like Synaptic Pruning Deficits," *Neuron* 83: 994–96. The discovery's implications for future treatments is the subject of an article by Pam Belluck, "Study Finds That Brains with Autism Fail to Trim Synapses as They Develop," *New York Times*, August 21, 2014.
129   *children can notice correlations*: Keil, "Science Starts Early," 1022.
130   *tasks do not require deep temporal depth perception*: Ball, "Cellular memory"; Balter "'Killjoys' Challenge Claims"; and Kagan, "The uniquely human in human nature" summarize some of the doubts raised about identifying human skills too closely with animal behaviors.
130   *"combine images and experiences"*: Balter, "Can Animals Envision the Future?"
132   *"daily pucker of blank anxiety"*: Bayley, *Elegy*, 63–64.
132–133   *"in humans, the ability to imagine future events"*: Shettleworth, "Planning for breakfast," 826. See also Daniel L. Schacter et al., "Remembering the past"
133   *it began in the First World War*: Although today well known in the Second World War, the strategy came into common, if less systematic, use in the first days of World War I. See Stéphane Audoin-Rouzeau and Annette Becker, *14–18: Understanding the Great War*, trans. Catherine Temerson (New York: Hill and Wang, 2002) chaps. 1–3.
133–134   *National and University Library of Bosnia and Herzegovina*: Of the 200,000 manuscripts, maps, personal archives, and pictures in the library's special collections, only 19,700 were recovered after the fire. See the International Federation of Library Associations and Institutions information at www.ifla.org/news/20-years-later-the-national-and-university-library-of-bosnia-and-herzegovina.
135   *"life is elsewhere"*: Kundera's book, *Life Is Elsewhere*, was published in 1973.

CHAPTER NINE: MASTERING MEMORY IN THE DIGITAL AGE

138   *"our brains are the ultimate general-purpose organ"*: Tattersall, *Masters*, 228.
138   *"Google Effects on Memory"*: Sparrow et al., "Google Effects on Memory." Tellingly, when people search for information they know is stored elsewhere, they recall where the information is stored.
139   *Twitter archive at the Library of Congress*: For news of the donation and terms of use, see "Update on the Twitter Archive at the Library of Congress," issued in January 2013. http://blogs.loc.gov/loc/2013/01/update-on-the-twitter-archive-at-the-library-of-congress/.

140   *"More important than technical approaches"*: McCarthy, "Reflections On," 1645.

142   *the results are easily gamed*: The Federal Trade Commission found Google "manipulated search results to favor its own services over rivals', even when they weren't most relevant for users." Rolfe Winkler and Brody Mullins, "How Google Skewed Search Results," *Wall Street Journal*, March 19, 2015.

143   *slow and fast thinking*: The ability to move beyond instinctual reactions and make conscious choices is what Daniel Kahneman discusses in his book *Thinking, Fast and Slow*, in which he draws a distinction between one system of thought that is reflexive, instinctual, intuitive, and fast and a more deliberative, effortful, conscious system of thinking.

144   *"Organizations are better"*: Ibid., 417–18.

144   *skills to control our personal information*: For a summary of essential digital good housekeeping, see Leslie Johnston, "Am I a Good Steward of My Own Digital Life?" The Signal, December 12, 2013. http://blogs.loc.gov/digitalpreservation/2013/12/am-i-a-good-steward-of-my-own-life/.

145   *conversion of ships' logs*: Schrope, "The Real Sea Change."

145–146   *collections of human and animal tissue samples*: The Armed Forces Institute of Pathology, for example, is the largest repository of its kind in the world. With nearly ninety million tissue samples, it includes "some of the most rare and difficult cases" in the history of medicine and is used by pathologists for diagnosis and categorization of disease. Despite the fact that it is a gold mine for genetic analysis and historical data about disease, it has been repeatedly subject to budget cuts by the Department of Defense and threatened with closure. Alison McCook, "Shelved."

146   *natural history museum specimen collections*: Kemp, "The endangered dead."

146   *the Avian Phylogenomics Project*: http://avian.genomics.cn/en/; Kress, "Valuing Collections," 1310.

146   *glass plate negatives*: Bhattacharjee, "Stars in Dusty Filing Cabinets."

146   *the first recording made in 1860*: The oldest known recording is a "phonautogram" made in France of a singer performing "Au Clair de Lune." When the creator, Édouard-Léon Scott de Martinville, made it in 1860, there was no playback equipment yet, so he never heard what he had recorded. To hear the restoration made by Carl Haber and Earl Cornell, see Jody Rosen, "Researchers Play Tune Recorded Before Edison," *New York Times*, March 27, 2008.

147   *imaging the tracks of subatomic particles*: Personal communication from Carl Haber.

147   *IRENE*: http://irene.lbl.gov/. Carl Haber's groundbreaking work on sound recovery is being developed for use in libraries and archives and

won him the MacArthur "genius award" in 2013. Lou Fancher, "Berkeley Lab's Carl Haber: A Genius in Our Midst," *Berkeleyside,* December 16, 2013.

147   *possibilities for the extraction of information*: In addition, mining databases about physical collections allows preservationists "to predict behaviors of collection materials and treatments based on characterization of these materials by non-invasive or optical methods." See the Library of Congress Preservation Directorate's website. http://www.loc.gov/preservation/outreach/tops/strlic_dahlberg/index.html.

150   *preserving creative content*: For fuller discussion about creative content, as well as scholarly, scientific, and open web content, see *Sustainable Economics.*

150   *exemptions for libraries and archives*: Section 108 of the U.S. copyright law grants to libraries and archives exemptions from copyright protection for preservation purposes. The institutions are allowed to make copies of items in their collections—books, photographs, LPs, and so forth—in order to preserve the content. This means that when a book still under copyright is full of acidic pages that are crumbling, the library or archives can make a microfilm, digital copy, or photocopy of the book in order to provide continued access to the content. This exemption applies to analog-based materials, and needs to be updated for the digital world, in which, technically speaking, every time someone brings up a digital file from the hard drive, it is a "copy." In March 2008, a report sponsored by the U.S. Copyright Office and the Library of Congress made recommendations to address the outstanding impediments to preserving born-digital content. See "The Section 108 Study Group Report" at www.section108.gov/docs/Sec108StudyGroup Report.pdf.

Most of the recommendations remain to be implemented. In a similar vein, another section of the copyright code allows libraries to purchase physical copies of copyrighted material and lend them for limited periods of time to their patrons. This first-sale doctrine also allows people to give their legally-purchased books to others and used bookstores to sell books. They are selling a copy of the book, not the underlying copyright of the content.

151   *economic and commercial value of content*: For a detailed analysis of economic models for digital data preservation, see *Sustainable Economics.*

152   *micropayments*: See, for example, Jaron Lanier's book *Who Owns the World?* New York: Simon & Schuster, 2013.

153   *September 11 Archive*: An overview of the September 11 Archive is available on the Roy Rosenzweig Center for History and New Media website at: http://chnm.gmu.edu/the-september-11-digital-archive/; and on the Library of Congress website at: http://lcweb2.loc.gov/diglib/lcwa/html/sept11/sept11-overview.html.

153 *Internet Archive*: The Internet Archive collects materials from the web through an automated system called harvesting. The method of the archive is to sample segments of the digital universe, taking captures of websites from time to time in a mode similar to telescopes dedicated to surveying deep space. The captures themselves are interactive: A site is navigable and its videos can be played, for example.

153–154 *international consortium to coordinate to coordinate digital collecting*: The alliance is the International Internet Preservation Consortium, http://netpreserve.org/.

CHAPTER TEN: BY MEMORY OF OURSELVES

160 *former NSA scientist said that staff*: Julia Angwin, "NSA Struggles to Make Sense of Surveillance Data." *Wall Street Journal*, December 25, 2013.

160 *"Our tendency to construct"*: Kahneman, *Thinking*, 218.

162 *"devised a system that encapsulates"*: "Long-term storage in DNA," 276.

163 *the limit to storage on silicon*: As *Nature* recently reported, Moore's law predicting the doubling of memory component storage every two years is now reaching its physical limits. "More from Moore," *Nature* 5, April 23, 2015, 408.

163 *"It makes Google's search problems"*: "The human brain produces in 30 seconds as much data as the Hubble Space Telescope has produced in its lifetime." Konrad Kording, quoted in Abbott, "Solving the Brain," 274. The BRAIN Initiative was funded in 2013 and allocated one hundred million dollars to start. The same year the EC announced its decade-long Human Brain Project with first year funding of €54M ($69M).

163 *one cubic millimeter of brain tissue generates two thousand terabytes of data*: For a quick and daunting overview of imaging parts of the brain, see Harvard University's Research Computing calculations at https://rc.fas.harvard.edu/case-studies/connections-in-the-brain/; and http://theastronomist.fieldofscience.com/2011/07/cubic-millimeter-of-your-brain.html.

165 *"20 petabytes of data"*: Information provided by Wendy Hanamura. Top 250 sites: www.alexa.com/siteinfo/archive.org.

165 *following links from Wikipedia*: www.alexa.com/topsites.

165 *study from 2013*: AlNoamany et al., "Who and What Links to the Internet Archive," 11.

167–168 *Long Now Foundation projects*: http://longnow.org/projects/.

168 *"there is nothing more deceptive"*: Doyle, "The Boscombe Valley Mystery" in *The Complete Original Illustrated Sherlock Holmes*.

168–169   *"to link the molecular"*: Dudai, "A Journey to Remember," 157.

169   *"This shift is usually described"*: Laughlin, *A Different Universe*, 76.

169   *the company spent twenty-one billion dollars on data centers*: www
.datacenterknowledge.com/archives/2013/09/17/google-has-spent-21-
billion-on-data-centers/. Now that Google Cloud Platform is competing
with Amazon Web Services, Microsoft, and IBM for corporate data
analysis and other cloud services, it is more open about its computing
capacity, including data back-up. See Quentin Hardy, "Google is Its
Own Secret Weapon in the Cloud," *New York Times*, June 1, 2015.

171–172   *"the great obstacle to good education"*: Jefferson in a letter to Nathaniel
Burwell, March 14, 1818, quoted in Sowerby, IV: 433 and using his
characteristically casual spelling and punctuation. "when this poison
infects the mind, it destroys it's [sic] tone, and revolts it against wholsome
reading. reason and fact, plain and unadorned are rejected. nothing can
engage attention unless dressed in all the figments of fancy; and nothing
so bedecked comes amiss. the result is a bloated imagination, sickly judgment, and disgust towards all the real business of life."

172   *a mere 20 percent of his books*: Sowerby lists 4,930 titles of which 757 fall
under the Fine Arts. (Her entry numbers are by title, not individual
volumes.) The Fine Arts category featured Jefferson's greatest passion,
building. He had books on architectural history, theory, and engineering, especially books by and about his idol Palladio, mixed in liberally
with treatises on building design and construction. He also included
collections on gardening, painting, sculpture; there were musical scores
(he played the violin and his daughter the keyboard); and a smattering of
literary forms from poetry and drama to didactic works. He collected
books on logic, from Aristotle to Condillac, compendia of exemplary
sermons, political speeches, essays on tragedy and comedy, illustrated
travel books—all categorized under Fine Arts. But the architect and
general contractor of Monticello was a practical futurist, giving pride of
place to his copious books on building and landscapes. Jefferson predicted
that the population would double every twenty years and Americans
would need to build lots of houses.

172   *"I know of nothing so pleasant"*: Lincoln in his Address before the
Wisconsin State Agricultural Society, Milwaukee, September 30, 1859.
http://riley.nal.usda.gov/nal_display/index.php?info_center=8&tax_level=
4&tax_subject=3&topic_id=1030&level3_id=6723&level4_id=11085.

173   *"is perfectible to a degree"*: "I am among those who think well of the
human character generally. I consider man as formed for society, and
endowed by nature with those dispositions which fit him for society. I
believe also, with Condorcet . . . that his mind is perfectible to a degree
of which we cannot as yet form any conception. It is impossible for a
man who takes a survey of what is already known, not to see what an

immensity in every branch of science yet remains to be discovered, & that too of articles to which our faculties seem adequate." Jefferson in a letter to William Green Munford, June 18, 1799.

173–174   *"Science fixes our attention"*: Jasanoff, "Technologies of humility," 33.

174   *"Science addresses what is true"*: Weinberg, "Reductionism Redux."

174   *"The historical process"*: Berlin, *Russian Thinkers*, 98.

Selected Sources

Abbott, Alison. 2009. "Brain imaging skewed." *Nature* 458: 1087.

———. 2013. "Solving the brain." *Nature* 499: 272–74.

Ainsworth, Claire. 2008. "Logbooks Record Weather's History." *Science* 322: 1629.

"All Hands on Deck." 2010. *Science* 330: 431.

AlNoamany, Yasmin, Ahmed AlSum, Michele C. Weigle, and Michael L. Nelson. 2013. "Who and What Links to the Internet Archive?" arXiv: 1309.4016v1 [cs.DL] September 16.

American Memory. http://memory.loc.gov.

American Treasures in the Library of Congress. 1997. Washington, D.C.: Library of Congress.

Angier, Nathalie. 2008. "Gut Instinct's Surprising Role as Precursor to Math." *New York Times*, September 16.

Appenzeller, Tim. 2013. "Old Masters." *Nature* 497: 302–4.

Arthur, W. Brian. 1999. "Complexity and the Economy." *Science* 284: 107–9.

Aubert, M., A. Brumm, M. Ramli, T. Sutikna, E. W. Saptomo, B. Hakim, M. J. Morwood, G. D. van den Bergh, L. Kinsley, and A. Dosseto. 2014. "Pleistocene cave art from Sulawesi, Indonesia." *Nature* 514: 223–27.

Augustine of Hippo. 2006. *Confessions.* 2nd ed. Translated by F. J. Sheed. Indianapolis, IN: Hackett Publishing.

Bacon, Francis. 2008. *Francis Bacon: The Major Works.* Edited by Brian Vickers. New York: Oxford University Press.

Bakewell, Sarah. 2010. *How to Live, or A Life of Montaigne in One Question and Twenty Attempts at an Answer.* New York: Other Press.

Ball, Philip. 2008. "Cellular memory hints at the origins of intelligence." *Nature* 451: 385.

Balter, Michael. 2006. "Radiocarbon Dating's Final Frontier." *Science* 313: 1560–63.

———. 2008. "Why We Are Different: Probing the Gap between Apes and Humans." *Science* 319: 404–5.

———. 2009. "Early Start for Human Art? Ochre May Revise Timeline." *Science* 323: 569.

———. 2009. "On the Origin of Art and Symbolism." *Science* 329: 709–11.

———. 2012. "'Killjoys' Challenge Claims of Clever Animals." *Science* 335: 1036–37.

———. 2013. "Can Animals Envision the Future? Scientists Spar Over New Data." *Science* 340: 909.

Battles, Matthew. 2003. *Library: An Unquiet History.* New York: W. W. Norton.

Bayley, John. 1999. *Elegy for Iris.* New York: Picador.

Beck, Melinda. 2010. "How to Tame Your Nightmares: Theories Teach Sleepers to Change the Ending of Their Dreams—or Even Take Flight." *Wall Street Journal,* July 20.

Bedini, Silvio A. 1990. *Thomas Jefferson: Statesmen of Science.* New York: Macmillan Publishing.

———. 2002. *Jefferson and Science.* Thomas Jefferson Foundation: Monticello Monograph Series.

Berlin, Isaiah. 1978. *Russian Thinkers.* Edited by Henry Hardy and Aileen Kelly. New York: Penguin Books.

Bernhard, Helen, Urs Fischbacher, and Ernst Fehr. 2006. "Parochial altruism in humans." *Nature* 442: 912–15.

Bhattacharjee, Yudhijit. 2008. "A Universe Past the Braking Point." *Science* 322: 1320–21.

———. 2009. "Stars in Dusty Filing Cabinets." *Science* 324: 460–61.

———. 2011. "Peering Back 13 Billion Years, Through a Gravitational Lens." *Science* 332: 522.

Bidney, David. 1947. "Human Nature and the Cultural Process." *American Anthropologist* 49: 375–99.

Bolhuis, Johan J. and Clyde D. L. Wynne. 2009. "Can evolution explain how minds work?" *Nature* 458: 832–33.

Bowler, Peter J. 2003. *Evolution: The History of an Idea.* 3rd ed. Berkeley: University of California Press.

Bowler, Peter J. and Iwan Rhys Morus. 2005. *Making Modern Science: A Historical Survey.* Chicago: University of Chicago Press.

Bowles, Samuel. 2008. "Conflict: altruism's midwife." *Nature* 456: 321–27.

———. 2009. "Did Warfare Among Ancestral Hunter-Gatherers Affect the Evolution of Social Behaviors?" *Science* 325: 1293–97.

Boyd, Robert and Peter J. Richerson. 2005. *The Origin and Evolution of Cultures.* New York: Oxford University Press.

Brook, Edward J. 2005. "Tiny Bubbles Tell All." *Science* 310: 1285–87.

Brown, Donald E. 1999. "Human Nature and History." *History and Theory* 38/4: 138–57.

——. 2004. "Human universals, human nature & human culture." *Daedalus* 2004/3: 47–53.

Brown, Mark J. S. 2011. "The trouble with bumblebees." *Nature* 469: 169–70.

Brumfiel, Geoff. 2011. "Down the petabyte highway." *Nature* 469: 282–83.

——. 2012. "Theorists feast on Higgs data." *Nature* 487: 281.

Buchanan, Mark. 2009. "Secret signals." *Nature* 547: 528–30.

Burrow, J. W. 2000. *The Crisis of Reason: European Thought, 1848–1914.* New Haven, CT: Yale University Press.

Buszáki, György. 2007. "The structure of consciousness." *Nature* 446: 267.

Canfora, Luciano. 1990. *The Vanished Library: A Wonder of the Western World.* Translated by Martin Ryle. Berkeley: University of California Press.

Carr, Annemarie Weyl. 2013. "Reading, Writing, and Books in Byzantium," in *Heaven and Earth, the Art of Byzantium from Greek Collections.* Edited by Anastasia Drandaki, Demetra Papanikola-Bakirtzi, and Anastasia Tourta. Athens: Benaki Museum.

Catalogue of the Library of Thomas Jefferson. 1952. Compiled and annotated by E. Millicent Sowerby. Washington, D.C.: Library of Congress.

Cho, Adrian. 2010. "What Shall We Do With the X-ray Laser?" *Science* 330: 1470–71.

——. 2011. "Have Physicists Already Glimpsed Particles of Dark Matter?" *Science* 331: 112–13.

Clark, Andy. 2008. *Supersizing the Mind: Embodiment, Action, and Cognitive Extension.* New York: Oxford University Press.

Clottes, Jean. 2008. *Cave Art.* London: Phaidon Press.

Cohen, I. Bernard. 1995. *Science and the Founding Fathers: Science in the Political Thought of Thomas Jefferson, Benjamin Franklin, John Adams and James Madison.* New York: W. W. Norton.

Cole, Michael, Karl Levitin, and Alexander Luria. 2010. *The Autobiography of Alexander Luria: A Dialogue with the Making of Mind.* New York: Psychology Press.

Coles, Peter. 2008. "Master of the Universe." *Science* 321: 1637.

Collins, Harry. 2009. "We cannot live by skepticism alone." *Nature* 458: 30–31.

Collins, James P. 2010. "Sailing on an Ocean of 0s and 1s." *Science* 327: 1455–56.

Conard, Nicholas J. 2009. "A female figurine from the basal Aurignacian of Hohle Fels Cave in southwestern Germany." *Nature* 459: 248–52.

Crease, Robert P. 2007. "Human distilleries." *Nature* 450: 350–51.

Culotta, Elizabeth. 2009. "On the Origin of Religion." *Science* 326: 783–87.

Curry, Andrew. 2006. "A Stone Age World Beneath the Baltic Sea." *Science* 314: 1533–35.

Custers, Ruud and Henk Aarts. 2010. "The Unconscious Will: How the Pursuit of Goals Operates Outside of Conscious Awareness." *Science* 329: 47–50.

Dalton, Rex. 2006. "Telling the time." *Nature* 444: 134–35.

———. 2009. "Ice-core researchers hope to chill out." *Nature* 460: 786–87.

Damasio, Antonio. 1999. *The Feeling of What Happens: Body and Emotion in the Making of Consciousness*. New York: Harcourt.

———. 2003. *Looking for Spinoza: Joy, Sorrow, and the Feeling Brain*. New York: Harcourt.

Danchin, Etienne, Luc-Alain Giradeau, Thomas J. Valone, and Richard H. Wagner. 2004. "Public Information: From Nosy Neighbors to Cultural Evolution." *Science* 305: 487–91.

Danielson, Dennis. 2001. "Scientist's birthright." *Nature* 410: 1031.

Danziger, Kurt. 2008. *Marking the Mind: A History of Memory*. Cambridge, UK: Cambridge University Press.

Darwin, Charles. 1964. *On the Origin of Species: A Facsimile of the First Edition*. Cambridge, MA: Harvard University Press.

———. 2011. *The Voyage of the* Beagle. New York: Modern Library.

"Data's shameful neglect." 2009. *Nature* 461: 45.

Davidson, Richard J. 2002. "Synaptic Substrates of the Implicit and Explicit Self." *Science* 296: 268.

De Martino, Benedetto, Dharsan Kumaran, Ben Seymour, and Raymond J. Dolan. 2006. "Frames, Biases, and Rational Decision-Making in the Human Brain." *Science* 313: 684–87.

Deacon, Terrence W. 1998. *The Symbolic Species: The Co-Evolution of Language and the Brain*. New York: W. W. Norton.

Dean, L. G., R. L. Kendal, S. J. Schapiro, B. Thierry, and K. N. Laland. 2012. "Identification of the Social Cognitive Processes Underlying Human Cumulative Culture." *Science* 335: 1114–18.

Dear, Peter. 2006. *The Intelligibility of Nature: How Science Makes Sense of the World*. Chicago: University of Chicago Press.

———. 2007. "The birth of science." *Nature* 446: 731.

Dolan, R. J. 2002. "Emotion, Cognition, and Behavior." *Science* 298: 1191–94.

Douglas, Mary. 2000. "Deep Thoughts on the Forbidden." *Science* 289: 2288.

Doyle, Arthur Conan. 1976. *The Complete Original Illustrated Sherlock Holmes*. Secaucus, NJ: Castle Books.

———. 1995. *The Adventures of Sherlock Holmes.* Norwalk, CT: Easton Press.

Drake, Nadia. 2014 "Cloud computing beckons scientists." *Nature* 509: 543–44.

Dudai, Yadin. 2006. "A journey to remember." *Nature* 441: 157–58.

Dudai, Yadin and Mary Carruthers. "The Janus face of Mnemosyne." *Nature* 434 (2005): 567.

Dyson, Freeman. 2009. "Leaping into the Grand Unknown." *New York Review of Books* 56/6: 59–61.

Ede, Siân. 2007. "An illusionary rival." *Nature* 448: 995–96.

Edelman, Gerald M. and Giulio Tononi. 2000. *A Universe of Consciousness: How Matter Becomes Imagination.* New York: Basic Books.

Edgerton, David. 2007. *The Shock of the Old: Technology and Global History Since 1900.* New York: Oxford University Press.

Eisenberg, Anne. 2000. "What's Next: An Electronic Circuit That Draws Its Inspiration from Life." *New York Times,* June 29.

Eisenstein, Elizabeth L. 1983. *The Printing Revolution in Early Modern Europe.* Cambridge, UK: Cambridge University Press.

Ellis, Joseph J. 1998. *American Sphinx: The Character of Thomas Jefferson.* New York: Vintage Books.

Evolution: The First Four Billion Years. 2009. Edited by Michael Ruse and Joseph Travis. Cambridge, MA: Belknap Press of Harvard University Press.

Febvre, Lucien and Henri-Jean Martin. 1997. *The Coming of the Book: The Impact of Printing 1450–1800.* Translated by David Gerard. London: Verso Classics.

Flanagan, Owen. 2011. "Knowing and feeling." *Nature* 469: 160–61.

Forth, Christopher E. 2009. "Imagining Our Ancient Future." *Science* 325: 677–78.

Frame, Donald M. 1984. *Montaigne: A Biography.* San Francisco: North Point Press.

Franklin, Benjamin. 1743. "A Proposal for Promoting Useful Knowledge in the British Plantations in America." National Humanities Center. http://nationalhumanitiescenter.org/pds/becomingamer/ideas/text4/amerphilsociety.pdf.

Frazzetto, Giovanni. 2012. "Powerful acts." *Nature* 482: 466–67.

Friedman, Jerome. 1999. "Creativity in Science." American Council of Learned Societies Occasional Paper 47.

From a Life of Physics. 1989. Contributions by H. A. Bethe, P. A. M. Dirac, W. Heisenberg, E. P. Wigner, O. Klein, and L. D. Landau (by E. M. Lifshitz). Singapore: World Scientific Publishing.

Gaddis, John Lewis. 2002. *The Landscape of History: How Historians Map the Past.* New York: Oxford University Press.

Galison, Peter. 2003. *Einstein's Clocks, Poincaré's Maps: Empires of Time.* New York: W. W. Norton.

Gazzaniga, Michael S. 2000. *The Mind's Past.* Berkeley: University of California Press.

———. 2005. *The Ethical Brain: The Science of Our Moral Dilemmas.* New York: HarperCollins.

Gee, Henry. 1999. *In Search of Deep Time: Beyond the Fossil Record to a New History of Life.* New York: Free Press.

Geertz, Clifford. 1983. *Local Knowledge: Further Essays in Interpretive Anthropology.* 3rd ed. New York: Basic Books.

Geography and Revolution. 2005. Edited by David N. Livingstone and Charles W. J. Withers. Chicago: University of Chicago Press.

Gervais, William G. and Ara Norenzayan. 2012. "Analytical Thinking Promotes Religious Disbelief." *Science* 336: 493–96.

Gibson, Ellen. 2011. "People become attached to smartphones." *Newsobserver.com*, August 1.

Giordano, Ralph, G. 2012. *The Architectural Ideology of Thomas Jefferson.* Jefferson, NC: McFarland.

Gleick, James. 2011. *The Information: A History. A Theory. A Flood.* New York: Pantheon Books.

Glimcher, Paul W. and Aldo Rustichini. 2004. "Neuronal Economics: The Consilience of Brain and Decision." *Science* 306: 447–52.

Goodrum, Charles A. 1980. *Treasures of the Library of Congress.* New York: Harry N. Abrams.

Goodrum, Charles A. and Helen W. Dalrymple. 1982. *Library of Congress.* 2nd ed. Boulder, CO: Westview Press.

Gould, Stephen Jay. 1982. "Nonmoral Nature." *Natural History* 91/2.

———. 1987. *Time's Arrow, Time's Cycle: Myth and Metaphor in the Discovery of Geological Time.* Cambridge, MA: Harvard University Press.

Greenblatt, Stephen. 2011. *The Swerve: How the World Became Modern.* New York: W. W. Norton.

Greenfield, Susan. 2000. *The Private Life of the Brain: Emotions, Consciousness, and the Secrets of the Self.* New York: John Wiley and Sons.

Gruber, Howard E. 1981. *Darwin on Man: A Psychological Study of Scientific Creativity.* 2nd ed. Chicago: University of Chicago Press.

Guise, Kevin, Karen Kelly, Jennifer Romanowski, Kai Vogeley, Steven M. Platek, Elizabeth Murray, and Julian Paul Keenan. 2007. "The Anatomical and Evolutionary Relationship between Self-awareness and Theory of Mind." *Human Nature* 18: 132–52.

Hacking, Ian. 2004. "Minding the Brain." *New York Review of Books* 51, June 24.

Haidt, Jonathan. 2007. "The New Synthesis in Moral Psychology." *Science* 316: 998–1002.

"Half Truths." 2008. *Science* 337: 270.

Hassabis, Demis, Dharshan Kumaran, Seralynne D. Vann, and Eleanor A. Maguire. "Patients with Hippocampal Amnesia Cannot Imagine New Experiences." 2007. *Proceedings of the National Academy of Sciences of the United States of America* 104: 1726–31.

Hauser, Oliver P., David G. Rand, Alexander Peysakhovich, and Martin A. Nowak. "Cooperating with the future." 2014. *Nature* 511: 220–23.

Hayden, Erika Check. 2009. "The other strand." *Nature* 457: 776–79.

Hedman, Matthew. 2007. *The Age of Everything: How Science Explores the Past.* Chicago: University of Chicago Press.

Heisenberg, Martin. 2009. "Is free will an allusion?" *Nature* 459: 164–65.

Hobson, J. Allan. 1999. *Consciousness.* New York: Scientific American Library.

Holtz, Robert Lee. 2007. "Most Science Studies Appear to Be Tainted by Sloppy Analysis." *Wall Street Journal*, September 14.

Hughes, Thomas P. 2004. *Human-Built World: How to Think About Technology and Culture.* Chicago: University of Chicago Press.

"Humanity and evolution." 2009. *Nature* 457: 763–64.

Hunt, Lynn. 2008. *Measuring Time, Making History.* Budapest-New York: Central European University Press.

Hutson, James H. 1998. *Religion and the Founding of the American Republic.* Washington, D.C.: Library of Congress.

Information and the Nature of Reality: From Physics to Metaphysics. 2010. Edited by Paul Davies and Niels Henrik Gregersen. New York: Cambridge University Press.

Ioannidis, John P. A. 2005. "Why Most Published Research Findings Are False." *PLoS Medicine* 2: 0696–701.

Jablonka, Eva and Marion J. Lamb. 2005. *Evolution in Four Dimensions: Genetic, Epigenetic, Behavioral, and Symbolic Variation in the History of Life.* Cambridge, MA: MIT Press.

Jackson, Stephen T. 2009. "Alexander von Humboldt and the General Physics of the Earth." *Science* 324: 596–97.

Jackson, Stephen T. and Richard J. Hobbs. 2009. "Ecological Restoration in the Light of Ecological History." *Science* 325: 567–68.

James, William. 1950. *The Principles of Psychology.* New York: Dover Publications.

——. 1987. *Writings 1902–1910.* New York: Library of America.

——. 1992. *Writings 1878–1899.* New York: Library of America.

——. 1994. *The Varieties of Religious Experience: A Study in Human Nature.* New York: Modern Library.

Jasanoff, Sheila. 2007. "Technologies of humility." *Nature* 450: 33.

Jeanneney, Jean-Noël. 2007. *Google and the Myth of Universal Knowledge.* Translated by Teresa Lavender Fagan. Chicago: University of Chicago Press.

Jefferson, Thomas. 1984. *Writings*. New York: Library of America.

John, Jeremy Leighton. 2009. "The future of saving our past." *Nature* 459: 775–76.

Kagan, Jerome. 2004. "The uniquely human in human nature." *Daedalus* 2004/3: 77–88.

———. 2006. *An Argument for Mind*. New Haven, CT: Yale University Press.

———. 2007. *What Is Emotion? History, Measures, and Meaning*. New Haven, CT: Yale University Press.

Kahneman, Daniel. 2011. *Thinking, Fast and Slow*. New York: Farrar, Straus and Giroux.

Kandel, Eric R. 2006. *In Search of Memory: The Emergence of a New Science of Mind*. New York: W. W. Norton.

———. 2012. *The Age of Insight: The Quest to Understand the Unconscious and Art, Mind, and Brain. From Vienna 1900 to the Present*. New York: Random House.

Kandel, Eric R. and Larry Squire. 2000. "Neuroscience: Breaking Down Scientific Barriers to the Study of the Brain." *Science* 290: 1113–20. DOI: 10.1126/science.290.5494.113.

Keil, Frank C. 2011. "Science Starts Early." *Science* 331: 1022–23.

Kemp, Christopher. 2015. "The endangered dead." *Nature* 518: 292–94.

Kerr, Richard A. 2005. "Ocean Flow Amplified, Not Triggered, Climate Change." *Science* 307: 1854.

King, Barbara J. 2011. "Is mental time travel what makes us human?" *Times Literary Supplement*, October 26.

Krauss, Lawrence M. 2011. "A Quantum Life." *Chronicle of Higher Education*, August 20.

Knight, Robert T. 2007. "Neural Networks Debunk Phrenology." *Science* 316: 1578–79.

Kolbert, Elizabeth. 2007. "Crash Course." *New Yorker*, May 14: 68–76.

Kress, John H. 2014. "Valuing Collections." *Science* 346: 1310.

Kurzban, Robert and H. Clark Barrett. 2012. "Origins of Cumulative Culture." *Science* 335: 1056–57.

Laland, Kevin N., Kim Sterelny, John Odling-Smee, William Hoppitt, and Tobias Uller. 2011. "Cause and Effect in Biology Revisited: Is Mayr's Proximate-Ultimate Dichotomy Still Useful?" *Science* 334: 1512–16.

Laughlin, Robert B. 2005. *A Different Universe: Reinventing Physics from the Bottom Down*. New York: Basic Books.

Lazer, David, Alex Pentland, Lada Adamic, Sinan Aral, Albert-László Barabási, Devon Brewer, Nicholas Christakis, Noshir Contractor, James Fowler, Myron Gutmann, Tony Jebara, Gary King, Michael Macy, Deb Roy, and Marshall Van Alstyne. 2009. "Computational Social Science." *Science* 323: 721–23.

Lee, Sue-Hyun, Jun-Hyeok Choi, Nuribalhae Lee, Hye-Ryeon Lee, Jae-Ick Kim, Nam-Kyung Yu, Sun-Lim Choi, Seung-Hee Lee, Hyoung Kim, and Bong-Kiun Kaang. 2008. "Synaptic Protein Degradation Underlies Destabilization of Retrieved Fear Memory." *Science* 319: 1253–56.

Lewis-Williams, David. 2002. *The Mind in the Cave: Consciousness and the Origins of Art*. London: Thames and Hudson.

Lewis-Williams, David and David Pearce. 2005. *Inside the Neolithic Mind: Consciousness, Cosmos, and the Realm of the Gods*. London: Thames and Hudson.

Lewontin, Richard. 2005. "The Wars over Evolution." *New York Review of Books*, October 20, 51–54.

Liberman, Nira and Yaacov Trope. 2008. "The Psychology of Transcending the Here and Now." *Science* 322: 1201–5.

Livingstone, David N. 2003. *Putting Science in Its Place: Geographies of Scientific Knowledge*. Chicago: University of Chicago Press.

Lloyd, Seth. 2007. "The quantum was quirky." *Nature* 450: 1167–68.

———. 2008. "Quantum Information Matters." *Science* 319: 1209–11.

"Long-term storage in DNA," *Nature* 518: 276.

Lubenov, Evgueniy V. and Athanassios G. Siapas. 2009. "Hippocampal theta oscillations are traveling waves." *Nature* 459: 534–39.

Luria, A. R. 1987. *The Mind of the Mnemonist: A Little Book about a Vast Memory*. Translated by Lynn Solotaroff. Cambridge, MA: Harvard University Press.

Lyons, Jonathan. 2013. *The Society for Useful Knowledge*. New York: Bloomsbury Press.

MacCulloch, Diarmaid. 2003. *The Reformation: A History*. New York: Penguin Books.

Macdougall, Doug. 2008. *Nature's Clocks: Scientists Measure the Age of Almost Everything*. Berkeley: University of California Press.

MacLeod, Christine. 2009. "The invention of heroes." *Nature* 460: 572–73.

Maier, Pauline. 1997. *American Scripture: The Making of the Declaration of Independence*. New York: Alfred A. Knopf.

Malakoff, David. 2000. "Does Science Drive the Productivity Train?" *Science* 289: 1274–76.

Mann, Adam. 2011. "The hunting of the dark." *Nature* 471: 433–35.

Marcus, Amy Dockser. 2012. "The Hard Science of Monkey Business." *Wall Street Journal*, March 30.

Margolit, Avishai. 2002. *The Ethics of Memory*. Cambridge, MA: Harvard University Press.

Martin, Edwin T. 1952. *Thomas Jefferson: Scientist*. New York: Collier Books.

Mason, Malia F., Michael I. Norton, John D. van Horn, Daniel M. Wegner, Scott T. Grafton, and C. Neil Macrae. 2007. "Wandering Minds: The

Default Network and Stimulus-Independent Thought." *Science* 315: 393–95.

Mayr, Ernst. 1982. *The Growth of Biological Thought: Diversity, Evolution, and Inheritance*. Cambridge, MA: Belknap Press of Harvard University Press.

———. 1992. "The Idea of Teleology." *Journal of the History of Ideas* 52/1: 117–35.

———. 1997. *This Is Biology: The Science of the Living World*. Cambridge, MA: Belknap Press of Harvard University Press.

———. 2004. *What Makes Biology Unique? Considerations on the Autonomy of a Scientific Discipline*. New York: Cambridge University Press.

McCarthy, James J. 2009. "Reflections On: Our Planet and Its Life, Origins, and Futures." *Science* 326: 1646–55.

McCook, Allison. 2011. "Shelved." *Nature* 47: 270–72.

Mehta, Mayank. 2007. "Fascinating rhythm." *Nature* 446: 27.

Mellars, Paul. 2009. "Origins of the female image." *Nature* 459: 176–77.

"Memory failure detected." 2011. *Times Higher Education*.

Menand, Louis. 2001. "A Marketplace of Ideas." American Council of Learned Societies Occasional Paper 49.

"Microscopic models." 2009. *Nature* 459: 615.

Miller, Greg. 2004. "Behavioral Neuroscience Uncaged." *Science* 306: 432–34.

———. 2007. "A Surprising Connection Between Memory and Imagination." *Science* 315: 312.

Miller, Jonathan F., Markus Neufang, Alec Solway, Armin Brandt, Michael Trippel, Irina Mader, Stefan Hefft, Max Merkow, Sean M. Polyn, Joshua Jacobs, Michael J. Kahana, Andreas Schulze-Bonhage. 2013. "Neural Activity in Human Hippocampal Formation Reveals the Spatial Context of Retrieved Memories." *Science* 342: 1111–14.

Milosz, Czeslaw. 1982. *Visions from San Francisco Bay*. Translated by Richard Lourie. New York: Farrar, Straus and Giroux.

———. 1983. *The Witness of Poetry*. Cambridge, MA: Harvard University Press.

Miyashita, Yasushi. 2004. "Cognitive Memory: Cellular and Network Machineries and Their Top-Down Control." *Science* 306: 435–40.

Montaigne, Michel de. 1943. *The Complete Works of Montaigne: Essays. Travel Journal. Letters*. Translated and edited by Donald M. Frame. Stanford: Stanford University Press.

Mulcahy, Nicholas J. and Josep Call. 2006. "Apes Save Tools for Future Use." *Science* 312: 1038–40.

Myin, Erik. 2010. "Unbounding the Mind." *Science* 330: 589–90.

Niebuhr, Reinhold. 1952. *The Irony of American History*. Chicago: University of Chicago Press.

Nisbet, Robert. 1980. *History of the Idea of Progress.* New York: Basic Books.

Nitz, Douglas. 2009. "The inside story on place cells." *Nature* 461: 889–90.

Nordling, Linda. 2010. "Researchers launch hunt for endangered data." *Nature* 468: 17.

Normile, Dennis. 2012. "Experiments Probe Languages' Origins and Development." *Science* 331: 408–11.

O'Connor, Ralph J. 2008. "Illuminating the Details of Deep Time." *Science* 321: 1447–48.

O'Hara, Kieron, Richard Morris, Nigel Shadbolt, Graham J. Hitch, Wendy Hall, and Neil Beagrie. 2006. "Memory for life: a review of the science and technology." *Journal of the Royal Society* 3: 351–65.

Oppezzo, Marily and Daniel L. Schwartz. 2014. "Give Your Ideas Some Legs: The Positive Effect of Walking on Creative Thinking." *Journal of Experimental Psychology: Learning, Memory, and Cognition* 40: 1142–52.

O'Reilly, Randall C. 2006. "Biologically Based Computational Models of High-Level Cognition." *Science* 314: 91–94.

Ostroff, Linnaea. 2011. "Recalling the future." *Nature* 474: 34.

Pääbo, Svante. 2014. *Neanderthal Man: In Search of Lost Genomes.* New York: Basic Books.

Padoa-Schioppa, Camillo and John A. Assad. 2006. "Neurons in the orbito-frontal cortex encode economic value." *Nature* 441: 223–26.

Page, F. W. 1895. "Our Library." *University of Virginia Alumni Bulletin* 1–2: 78–85.

Palmer, Linda and Gary Lynch. 2010. "A Kantian View of Space." *Science* 328: 1487–88.

Pennisi, Elizabeth. 2009. "On the Origin of Cooperation." *Science* 325: 1196–99.

Pesic, Peter. 1999. "Wrestling With Proteus: Francis Bacon and the 'Torture' of Nature." *Isis* 90/1:81–94.

Pico, Richard M. 2002. *Consciousness in Four Dimensions: Biological Relativity and the Origins of Thought.* New York: McGraw-Hill.

Pierpont, Claudia Roth. 2004. "The Measure of America." *New Yorker,* March 8, 48–63.

Plato. 1925. *Phaedrus: Plato in Twelve Volumes.* Vol. 9. Translated by Harold N. Fowler. Cambridge, MA: Harvard University Press; London: William Heinemann.

Pollack, Andrew. 2011. "DNA Sequencing Caught in Deluge of Data." *New York Times,* November 30.

Pollack, Robert. 1999. *The Missing Moment: How the Unconscious Shapes Modern Science.* Boston: Houghton Mifflin.

Powell, Adam, Stephen Shannon, and Mark G. Thomas. 2009. "Late Pleistocene Demography and the Appearance of Modern Human Behavior." *Science* 324: 1298–1301.

Preserving Our Digital Heritage: Plan for the National Digital Information Infrastructure and Preservation Program. 2002. Washington, D.C.: Library of Congress.

Pronin, Emily. 2008. "How We See Ourselves and How We See Others." *Science* 320: 1177–80.

Quammen, David. 2006. *The Reluctant Mr. Darwin*. New York: W. W. Norton.

Raddick, M. Jordan and Alexander S. Szalay. 2012. "The Universe Online." *Science* 329: 1028–29.

Raichle, Marcus E. 2006. "The Brain's Dark Energy." *Science* 314: 1249–50.

Ramirez, Steve, Xu Liu, Pei-Ann Lin, Junghyup Suh, Michele Pignatelli, Roger L. Redondo, Tomás J. Ryan, and Susumu Tonegawa. 2013. "Creating a False Memory in the Hippocampus." *Science* 341: 387–91.

Reich, Eugenie Samuel. 2011. "Tevatron's legacy set to disappear." *Nature* 474: 16–17.

Richerson, Peter J. and Robert Boyd. 2005. *Not by Genes Alone. How Culture Transforms Human Evolution*. Chicago: University of Chicago Press.

Richet, Pascal. 2007. *A Natural History of Time*. Translated by John Venerella. Chicago: University of Chicago Press.

Ricoeur, Paul. 2004. *Memory, History, Forgetting*. Translated by Kathleen Blamey and David Pellauer. Chicago: University of Chicago Press.

Robinson, Andrew. 2008. "A century of puzzling." *Nature* 453: 990–91.

Rodden, Appletree. 2011. "What makes us laugh." *Nature* 473: 450.

Roediger, H. L. III and K. A. DeSoto. 2014. "Forgetting the Presidents." *Science* 346: 1106–9.

Ross, Sydney. 1962. "Scientist: the Story of Word." *Annals of Science* 18, 2: 65–85.

Rossi, Paolo. 1984. *The Dark Abyss of Time: The History of the Earth and the History of Nations from Hooke to Vico*. Translated by Lydia G. Cochrane. Chicago: University of Chicago Press.

———. 2000. *Logic and the Art of Memory: The Quest for a Universal Language*. Translated by Stephen Clucas. Chicago: University of Chicago Press.

Rudwick, Martin J. S. 2005. *Bursting the Limits of Time: The Reconstruction of Geohistory in the Age of Revolution*. Chicago: University of Chicago Press.

———. 2008. *Worlds Before Adam. The Reconstruction of Geohistory in the Age of Reform*. Chicago: University of Chicago Press.

[Rumsey], Abby Smith. 2003. "Authenticity and Artifact: When Is a Watch Not a Watch?" *Library Trends* 52/1: 172–82.

Sahakian, Barbara, Andrew Lawrence, Luke Clark, and Jamie Nicole Labuzetta. 2008. "The innovative brain." *Nature* 456: 168–69.

Sarewitz, Daniel. 2012. "Beware the creeping cracks of bias." *Nature* 485: 149.

Schacter, Daniel L. and Donna Rose Addis. 2007. "The ghosts of past and future." *Nature* 445: 27.

Schacter, D. L., D. R. Addis, and R. L. Buckner. 2007. "Remembering the past to imagine the future: the prospective brain." *Nature Reviews. Neuroscience* 9: 675–661.

Schnabel, Jim. 2009. "Rethinking rehab." *Nature* 458: 25–27.

Schrope, Mark. 2006. "The real sea change." *Nature* 443: 622–24.

Schubert, Stephen Blake. 1993. "The Oriental Origins of the Alexandrian Library." *Libri* 43/2: 142–72.

Science of Memory: Concepts. 2007. Edited by Henry L. Roediger III, Yadin Dudai, and Susan M. Fitzpatrick. New York: Oxford University Press.

Shapin, Steven. 1996. *The Scientific Revolution.* Chicago: University of Chicago Press.

———. 2001a. "How to Be Antiscientific." In *The One Culture? A Conversation about Science.* Edited by J. A. Labinger and Harry Collins. Chicago: University of Chicago Press.

———. 2001b. "Proverbial Economies: How an Understanding of Some Linguistic and Social Features of Common Sense Can Throw Light on More Prestigious Bodies of Knowledge, Science for Example." *Social Studies of Science* 31/5: 713–69.

Sharp, Phillip A. 2014. "Meeting Global Challenges: Discovery and Innovation through Convergence." *Science* 348: 1468–71.

Shaw, Jonathan and Jennifer Carling. 2008. "Eye on the Universe." *Harvard Magazine* July–August: 30–35.

Shettleworth, Sara J. 2007. "Planning for breakfast." *Nature* 445: 825–26.

Shryock, Andrew and Daniel Lord Smail. 2011. *Deep History. The Architecture of Past and Present.* Berkeley: University of California Press.

———. 2013. "History and the 'Pre.'" *American Historical Review* 118: 709–37.

Sibum. H. Otto. 2004. "What Kind of Science Is Experimental Physics?" *Science* 306: 60–61.

Skoyles, John R. 2010. "Optimizing Scientific Reasoning." *Science* 330: 1477.

Smail, Daniel Lord. 2008. *On Deep History and the Brain.* Berkeley: University of California Press.

Smith, Eliot R. and Diane M. Mackie. 2009. "Surprising Emotions." *Science* 323: 215–16.

Smolin, Lee. 2010. "Space-time turn around." *Nature* 467: 1034–35.

Snyder, Laura J. 2011. *The Philosophical Breakfast Club.* New York: Broadway Books.

Sparrow, Betsy, Jenny Liu, and Daniel M. Wegner. 2011. "Google Effects on Memory. Cognitive Consequences of Having Information at Our Fingertips." *Science* 333: 776–78.

Squire, Larry R. and Eric R. Kandel. 1999. *Memory: From Mind to Molecules.* New York: Scientific American Library.

Stafford, Ned. 2010. "Science in the digital age." *Nature* 467: S19–S21.

Stirling, Andy. 2010. "Keep it complex." *Nature* 468: 1029–31.

Suddendorf, Thomas. 2006. "Foresight and Evolution of the Human Mind." *Science* 312: 1006–7.

"Sustainable Economics for a Digital Planet: Ensuring Long-term Access to Digital Information. Final Report of Blue Ribbon Task Force on Sustainable Digital Preservation and Access." 2010.

Szathmáry, Eőrs and Szabolcs Számadó. 2008. "Language: A Social History of Words." *Nature* 456: 40–41.

Talmi, Deborah and Chris Frith. 2007. "Feeling great about doing right." *Nature* 446: 865–66.

Tattersall, Ian. 2012. *Masters of the Planet: The Search for Our Human Origins.* New York: Palgrave McMillan.

Tenenbaum, Joshua B., Charles Kemp, Thomas L. Griffiths, and Noah D. Goodman. 2011. "How to Grow a Mind: Statistics, Structure, and Abstraction." *Science* 331: 1279–85.

The 1812 Catalogue of the Library of Congress. A Facsimile. 1982. Washington, D.C.: Library of Congress.

The Anatomy of Memory: An Anthology. 1996. Edited by James McConkey. New York: Oxford University Press.

"The Evidence in Hand: Report of the Task Force on the Artifact in Library Collections." 2001. Washington, D.C.: Council on Library and Information Resources.

"The human epoch." 2011. *Nature* 473: 254.

The Jefferson Bicentennial 1743–1943: A Catalogue of the Exhibitions in the Library of Congress. 1943. Washington, D.C.: Library of Congress.

The Legacy of Isaiah Berlin. 2001. Edited by Mark Lilla, Ronald Dworkin, and Robert Silvers. New York: New York Review of Books.

"The map in your head." 2009. *Nature* 459: 477.

Thomas Jefferson's Library: A Catalog with the Entries in His Own Order. 1989. Edited by James Gilreath and Doulas L. Wilson. Washington, D.C.: Library of Congress.

Thomas Jefferson's Poplar Forest: A Private Place. 2002. Corporation for Jefferson's Poplar Forest.

Thompson, Scott M. and Hayley A. Mattison. 2009. "Secret of synapse specificity." *Nature* 458: 296–97.

Thomson, Keith. 2012. *Jefferson's Shadow: The Story of His Science.* New Haven: Yale University Press.

Thurman, Judith. 2008. "First Impressions: What Does the World's Oldest Art Say About Us?" *New Yorker*, June 23, 58–67.

Turner, Fred. 2006. *From Counterculture to Cyberculture: Stewart Brand, the Whole Earth Network, and the Rise of Digital Utopianism.* Chicago: University of Chicago Press.

———. 2009. "Capturing digital lives." *Nature* 461: 1206–8.

Ulam, S. M. 1991. *Adventures of a Mathematician.* Berkeley: University of California Press.

Underwood, Emily. 2014. "Brain's GPS Finds Top Honor." *Science* 346: 149.

Velez, Juan-Pablo. 2011. "An Unusual Library Finds a Home." *New York Times*, November 12.

"Vive la révolution." 2011. *Nature* 469: 443.

Vogel, Gretchen. 2004. "Behavioral Evolution: The Evolution of the Golden Rule." *Science* 303: 1128–31.

Von Baeyer, Hans Christian. 2004. *Information: The New Language of Science.* Cambridge, MA: Harvard University Press.

von Neumann, John. 2002. *The Computer and the Brain.* 2nd ed. New Haven, CT: Yale University Press.

Weinberg, Steven. 1987. "Newtonianism, reductionism, and the art of congressional testimony." *Nature* 330: 433–37.

———. 1995. "Reductionism Redux." *New York Review of Books*, 42/15.

Whiten, Andrew. 2014. "Incipient tradition in wild chimpanzees." *Nature* 514: 178–79.

Wigner, Eugene. 1960. "The Unreasonable Effectiveness of Mathematics in Natural Sciences." *Communications in Pure and Applied Mathematics* 13/1.

———. 1992. *The Recollections of Eugene P. Wigner as Told to Andrew Szanton.* New York: Basic Books.

Willis, K. J. and H. J. B. Birks. 2006. "What Is Natural? The Need for a Long-Term Perspective and Biodiversity Conservation." *Science* 314: 1261–65.

Wilson, Douglas L. 1984. "Sowerby Revisited: The Unfinished Catalogue of Thomas Jefferson's Library." *William and Mary Quarterly* 36: 503–23.

Wilson, Edward O. 1998. *Consilience: The Unity of Knowledge.* New York: Alfred A. Knopf.

———. 2012. "On the Origins of the Arts." *Harvard Magazine*, May–June.

Wolf, Maryanne. 2007. *Proust and the Squid: The Story and Science of the Reading Brain.* New York: Harper Perennial.

Wood, Gordon S. 2006. *Revolutionary Characters: What Made the Founders Different.* New York: Penguin Books.

Yang, Guang, and Cora Sau Wan Lai, Joseph Cichon, Lei Ma, Wei Li, Wen-biao Gan. 2014. "Sleep promotes branch-specific formation of dendritic spines after learning." *Science* 344: 1173–78.

Yates, Frances A. 2001. *The Art of Memory*. London: Pimlico.

Zimmer, Carl. 2009. "On the Origin of Life on Earth." *Science* 323: 198–99.

——. 2011. "Nonfiction: Nabokov Theory on Butterfly Evolution is Vindicated." *New York Times*, January 25.

ILLUSTRATION CREDITS

1 Rough Draft of Declaration of Independence. Courtesy of the Library of Congress.
2 Chauvet Cave. Getty Images.
3 Pushmi-pullyu from *The Story of Doctor Dolittle*. Abby Smith Rumsey.
4 Cuneiform from library of Ashurbanipal. © Trustees of the British Museum.
5 Giant Bible of Mainz and Gutenberg Bible. Courtesy of the Library of Congress.
6 Charles Willson Peale Portrait of Thomas Jefferson. Courtesy of Independence National Historical Park.
7 Jefferson catalogue, 1815. Courtesy of the Library of Congress.
8 Sainte-Geneviève Library. © La Collection/Jean-Claude N'Diaye. Delivered on September 11, 2015.
9 F. Quesnel drawing of Michel de Montaigne, ca. 1588. Reprinted with permission from the Montaigne Studies website.
10 Alexander Romanovich Luria. Reprinted with permission from www.luria.ucsd.edu.
11 *Idealer Durchschnitt eines Theils Der Erdrinde*, by Heinrich Berghaus, 1841. David Rumsey Map Collection, www.davidrumsey.com.
12 Carl Haber and IRENE. © John D. and Catherine T. MacArthur Foundation. Licensed under Creative Commons, http://creativecommons.org/licenses/by/4.0/legalcode.
13 Library of Congress, 1897. Courtesy of the Library of Congress.
14 Brewster Kahle of the Internet Archive. AP Images/Ben Margot.

INDEX

Abby Smith Rumsey is a historian who writes about how ideas and information technologies shape perceptions of history, of time, and of personal and cultural identity. Trained at Harvard as a Russian scholar, she has worked in Soviet-era archives, spent a decade at the Library of Congress, and has consulted on digital collecting and curation, intellectual property issues, and the economics of digital information for a variety of universities and the National Science Foundation. She lives in San Francisco.